GEOMETRY

PURE AND APPLIED MATHEMATICS

A Program of Monographs, Textbooks, and Lecture Notes

EXECUTIVE EDITORS

Earl J. Taft
Rutgers University
New Brunswick, New Jersey

Zuhair Nashed
University of Delaware
Newark, Delaware

CHAIRMEN OF THE EDITORIAL BOARD

S. Kobayashi
University of California, Berkeley
Berkeley, California

Edwin Hewitt
University of Washington
Seattle, Washington

EDITORIAL BOARD

M. S. Baouendi
University of California, San Diego

Donald Passman
University of Wisconsin–Madison

Jack K. Hale
Georgia Institute of Technology

Fred S. Roberts
Rutgers University

Marvin Marcus
University of California,
Santa Barbara

Gian-Carlo Rota
Massachusetts Institute of
Technology

W. S. Massey
Yale University

David L. Russell
Virginia Polytechnic Institute
and State University

Leopoldo Nachbin
Centro Brasileiro de Pesquisas Físicas
and University of Rochester

Jane Cronin Scanlon
Rutgers University

Anil Nerode
Cornell University

Walter Schempp
Universität Siegen

Mark Teply
University of Wisconsin–Milwaukee

MONOGRAPHS AND TEXTBOOKS IN PURE AND APPLIED MATHEMATICS

1. *K. Yano*, Integral Formulas in Riemannian Geometry (1970)
2. *S. Kobayashi*, Hyperbolic Manifolds and Holomorphic Mappings (1970)
3. *V. S. Vladimirov*, Equations of Mathematical Physics (A. Jeffrey, editor; A. Littlewood, translator) (1970)
4. *B. N. Pshenichnyi*, Necessary Conditions for an Extremum (L. Neustadt, translation editor; K. Makowski, translator) (1971)
5. *L. Narici, E. Beckenstein, and G. Bachman*, Functional Analysis and Valuation Theory (1971)
6. *S. S. Passman*, Infinite Group Rings (1971)
7. *L. Dornhoff*, Group Representation Theory (in two parts). Part A: Ordinary Representation Theory. Part B: Modular Representation Theory (1971, 1972)
8. *W. Boothby and G. L. Weiss (eds.)*, Symmetric Spaces: Short Courses Presented at Washington University (1972)
9. *Y. Matsushima*, Differentiable Manifolds (E. T. Kobayashi, translator) (1972)
10. *L. E. Ward, Jr.*, Topology: An Outline for a First Course (1972)
11. *A. Babakhanian*, Cohomological Methods in Group Theory (1972)
12. *R. Gilmer*, Multiplicative Ideal Theory (1972)
13. *J. Yeh*, Stochastic Processes and the Wiener Integral (1973)
14. *J. Barros-Neto*, Introduction to the Theory of Distributions (1973)
15. *R. Larsen*, Functional Analysis: An Introduction (1973)
16. *K. Yano and S. Ishihara*, Tangent and Cotangent Bundles: Differential Geometry (1973)
17. *C. Procesi*, Rings with Polynomial Identities (1973)
18. *R. Hermann*, Geometry, Physics, and Systems (1973)
19. *N. R. Wallach*, Harmonic Analysis on Homogeneous Spaces (1973)
20. *J. Dieudonné*, Introduction to the Theory of Formal Groups (1973)
21. *I. Vaisman*, Cohomology and Differential Forms (1973)
22. *B.-Y. Chen*, Geometry of Submanifolds (1973)
23. *M. Marcus*, Finite Dimensional Multilinear Algebra (in two parts) (1973, 1975)
24. *R. Larsen*, Banach Algebras: An Introduction (1973)
25. *R. O. Kujala and A. L. Vitter (eds.)*, Value Distribution Theory: Part A; Part B: Deficit and Bezout Estimates by Wilhelm Stoll (1973)
26. *K. B. Stolarsky*, Algebraic Numbers and Diophantine Approximation (1974)
27. *A. R. Magid*, The Separable Galois Theory of Commutative Rings (1974)
28. *B. R. McDonald*, Finite Rings with Identity (1974)
29. *J. Satake*, Linear Algebra (S. Koh, T. A. Akiba, and S. Ihara, translators) (1975)
30. *J. S. Golan*, Localization of Noncommutative Rings (1975)
31. *G. Klambauer*, Mathematical Analysis (1975)
32. *M. K. Agoston*, Algebraic Topology: A First Course (1976)
33. *K. R. Goodearl*, Ring Theory: Nonsingular Rings and Modules (1976)
34. *L. E. Mansfield*, Linear Algebra with Geometric Applications: Selected Topics (1976)
35. *N. J. Pullman*, Matrix Theory and Its Applications (1976)
36. *B. R. McDonald*, Geometric Algebra Over Local Rings (1976)
37. *C. W. Groetsch*, Generalized Inverses of Linear Operators: Representation and Approximation (1977)
38. *J. E. Kuczkowski and J. L. Gersting*, Abstract Algebra: A First Look (1977)
39. *C. O. Christenson and W. L. Voxman*, Aspects of Topology (1977)
40. *M. Nagata*, Field Theory (1977)

41. *R. L. Long,* Algebraic Number Theory (1977)
42. *W. F. Pfeffer,* Integrals and Measures (1977)
43. *R. L. Wheeden and A. Zygmund,* Measure and Integral: An Introduction to Real Analysis (1977)
44. *J. H. Curtiss,* Introduction to Functions of a Complex Variable (1978)
45. *K. Hrbacek and T. Jech,* Introduction to Set Theory (1978)
46. *W. S. Massey,* Homology and Cohomology Theory (1978)
47. *M. Marcus,* Introduction to Modern Algebra (1978)
48. *E. C. Young,* Vector and Tensor Analysis (1978)
49. *S. B. Nadler, Jr.,* Hyperspaces of Sets (1978)
50. *S. K. Segal,* Topics in Group Kings (1978)
51. *A. C. M. van Rooij,* Non-Archimedean Functional Analysis (1978)
52. *L. Corwin and R. Szczarba,* Calculus in Vector Spaces (1979)
53. *C. Sadosky,* Interpolation of Operators and Singular Integrals: An Introduction to Harmonic Analysis (1979)
54. *J. Cronin,* Differential Equations: Introduction and Quantitative Theory (1980)
55. *C. W. Groetsch,* Elements of Applicable Functional Analysis (1980)
56. *I. Vaisman,* Foundations of Three-Dimensional Euclidean Geometry (1980)
57. *H. I. Freedan,* Deterministic Mathematical Models in Population Ecology (1980)
58. *S. B. Chae,* Lebesgue Integration (1980)
59. *C. S. Rees, S. M. Shah, and C. V. Stanojević,* Theory and Applications of Fourier Analysis (1981)
60. *L. Nachbin,* Introduction to Functional Analysis: Banach Spaces and Differential Calculus (R. M. Aron, translator) (1981)
61. *G. Orzech and M. Orzech,* Plane Algebraic Curves: An Introduction Via Valuations (1981)
62. *R. Johnsonbaugh and W. E. Pfaffenberger,* Foundations of Mathematical Analysis (1981)
63. *W. L. Voxman and R. H. Goetschel,* Advanced Calculus: An Introduction to Modern Analysis (1981)
64. *L. J. Corwin and R. H. Szcarba,* Multivariable Calculus (1982)
65. *V. I. Istrățescu,* Introduction to Linear Operator Theory (1981)
66. *R. D. Järvinen,* Finite and Infinite Dimensional Linear Spaces: A Comparative Study in Algebraic and Analytic Settings (1981)
67. *J. K. Beem and P. E. Ehrlich,* Global Lorentzian Geometry (1981)
68. *D. L. Armacost,* The Structure of Locally Compact Abelian Groups (1981)
69. *J. W. Brewer and M. K. Smith, eds.,* Emily Noether: A Tribute to Her Life and Work (1981)
70. *K. H. Kim,* Boolean Matrix Theory and Applications (1982)
71. *T. W. Wieting,* The Mathematical Theory of Chromatic Plane Ornaments (1982)
72. *D. B. Gauld,* Differential Topology: An Introduction (1982)
73. *R. L. Faber,* Foundations of Euclidean and Non-Euclidean Geometry (1983)
74. *M. Carmeli,* Statistical Theory and Random Matrices (1983)
75. *J. H. Carruth, J. A. Hildebrant, and R. J. Koch,* The Theory of Topological Semigroups (1983)
76. *R. L. Faber,* Differential Geometry and Relativity Theory: An Introduction (1983)
77. *S. Barnett,* Polynomials and Linear Control Systems (1983)
78. *G. Karpilovsky,* Commutative Group Algebras (1983)
79. *F. Van Oystaeyen and A. Verschoren,* Relative Invariants of Rings: The Commutative Theory (1983)
80. *I. Vaisman,* A First Course in Differential Geometry (1984)
81. *G. W. Swan,* Applications of Optimal Control Theory in Biomedicine (1984)
82. *T. Petrie and J. D. Randall,* Transformation Groups on Manifolds (1984)

83. K. Goebel and S. Reich, Uniform Convexity, Hyperbolic Geometry, and Nonexpansive Mappings (1984)
84. T. Albu and C. Năstăsescu, Relative Finiteness in Module Theory (1984)
85. K. Hrbacek and T. Jech, Introduction to Set Theory: Second Edition, Revised and Expanded (1984)
86. F. Van Oystaeyen and A. Verschoren, Relative Invariants of Rings: The Noncommutative Theory (1984)
87. B. R. McDonald, Linear Algebra Over Commutative Rings (1984)
88. M. Namba, Geometry of Projective Algebraic Curves (1984)
89. G. F. Webb, Theory of Nonlinear Age-Dependent Population Dynamics (1985)
90. M. R. Bremner, R. V. Moody, and J. Patera, Tables of Dominant Weight Multiplicities for Representations of Simple Lie Algebras (1985)
91. A. E. Fekete, Real Linear Algebra (1985)
92. S. B. Chae, Holomorphy and Calculus in Normed Spaces (1985)
93. A. J. Jerri, Introduction to Integral Equations with Applications (1985)
94. G. Karpilovsky, Projective Representations of Finite Groups (1985)
95. L. Narici and E. Beckenstein, Topological Vector Spaces (1985)
96. J. Weeks, The Shape of Space: How to Visualize Surfaces and Three-Dimensional Manifolds (1985)
97. P. R. Gribik and K. O. Kortanek, Extremal Methods of Operations Research (1985)
98. J.-A. Chao and W. A. Woyczynski, eds., Probability Theory and Harmonic Analysis (1986)
99. G. D. Crown, M. H. Fenrick, and R. J. Valenza, Abstract Algebra (1986)
100. J. H. Carruth, J. A. Hildebrant, and R. J. Koch, The Theory of Topological Semigroups, Volume 2 (1986)
101. R. S. Doran and V. A. Belfi, Characterizations of C*-Algebras: The Gelfand-Naimark Theorems (1986)
102. M. W. Jeter, Mathematical Programming: An Introduction to Optimization (1986)
103. M. Altman, A Unified Theory of Nonlinear Operator and Evolution Equations with Applications: A New Approach to Nonlinear Partial Differential Equations (1986)
104. A. Verschoren, Relative Invariants of Sheaves (1987)
105. R. A. Usmani, Applied Linear Algebra (1987)
106. P. Blass and J. Lang, Zariski Surfaces and Differential Equations in Characteristic p > 0 (1987)
107. J. A. Reneke, R. E. Fennell, and R. B. Minton, Structured Hereditary Systems (1987)
108. H. Busemann and B. B. Phadke, Spaces with Distinguished Geodesics (1987)
109. R. Harte, Invertibility and Singularity for Bounded Linear Operators (1988)
110. G. S. Ladde, V. Lakshmikantham, and B. G. Zhang, Oscillation Theory of Differential Equations with Deviating Arguments (1987)
111. L. Dudkin, I. Rabinovich, and I. Vakhutinsky, Iterative Aggregation Theory: Mathematical Methods of Coordinating Detailed and Aggregate Problems in Large Control Systems (1987)
112. T. Okubo, Differential Geometry (1987)
113. D. L. Stancl and M. L. Stancl, Real Analysis with Point-Set Topology (1987)
114. T. C. Gard, Introduction to Stochastic Differential Equations (1988)
115. S. S. Abhyankar, Enumerative Combinatorics of Young Tableaux (1988)
116. H. Strade and R. Farnsteiner, Modular Lie Algebras and Their Representations (1988)
117. J. A. Huckaba, Commutative Rings with Zero Divisors (1988)
118. W. D. Wallis, Combinatorial Designs (1988)
119. W. Więsław, Topological Fields (1988)
120. G. Karpilovsky, Field Theory: Classical Foundations and Multiplicative Groups (1988)
121. S. Caenepeel and F. Van Oystaeyen, Brauer Groups and the Cohomology of Graded Rings (1989)

122. W. Kozlowski, Modular Function Spaces (1988)
123. E. Lowen-Colebunders, Function Classes of Cauchy Continuous Maps (1989)
124. M. Pavel, Fundamentals of Pattern Recognition (1989)
125. V. Lakshmikantham, S. Leela, and A. A. Martynyuk, Stability Analysis of Nonlinear Systems (1989)
126. R. Sivaramakrishnan, The Classical Theory of Arithmetic Functions (1989)
127. N. A. Watson, Parabolic Equations on an Infinite Strip (1989)
128. K. J. Hastings, Introduction to the Mathematics of Operations Research (1989)
129. B. Fine, Algebraic Theory of the Bianchi Groups (1989)
130. D. N. Dikranjan, I. R. Prodanov, and L. N. Stoyanov, Topological Groups: Characters, Dualities, and Minimal Group Topologies (1989)
131. J. C. Morgan II, Point Set Theory (1990)
132. P. Biler and A. Witkowski, Problems in Mathematical Analysis (1990)
133. H. J. Sussmann, Nonlinear Controllability and Optimal Control (1990)
134. J.-P. Florens, M. Mouchart, and J. M. Rolin, Elements of Bayesian Statistics (1990)
135. N. Shell, Topological Fields and Near Valuations (1990)
136. B. F. Doolin and C. F. Martin, Introduction to Differential Geometry for Engineers (1990)
137. S. S. Holland, Jr., Applied Analysis by the Hilbert Space Method (1990)
138. J. Okniński, Semigroup Algebras (1990)
139. K. Zhu, Operator Theory in Function Spaces (1990)
140. G. B. Price, An Introduction to Multicomplex Spaces and Functions (1991)
141. R. B. Darst, Introduction to Linear Programming: Applications and Extensions (1991)
142. P. L. Sachdev, Nonlinear Ordinary Differential Equations and Their Applications (1991)
143. T. Husain, Orthogonal Schauder Bases (1991)
144. J. Foran, Fundamentals of Real Analysis (1991)
145. W. C. Brown, Matrices and Vector Spaces (1991)
146. M. M. Rao and Z. D. Ren, Theory of Orlicz Spaces (1991)
147. J. S. Golan and T. Head, Modules and the Structures of Rings: A Primer (1991)
148. C. Small, Arithmetic of Finite Fields (1991)
149. K. Yang, Complex Algebraic Geometry: An Introduction to Curves and Surfaces (1991)
150. D. G. Hoffman, D. A. Leonard, C. C. Lindner, K. T. Phelps, C. A. Rodger, and J. R. Wall, Coding Theory: The Essentials (1991)
151. M. O. González, Classical Complex Analysis (1992)
152. M. O. González, Complex Analysis: Selected Topics (1992)
153. L. W. Baggett, Functional Analysis: A Primer (1992)
154. M. Sniedovich, Dynamic Programming (1992)
155. R. P. Agarwal, Difference Equations and Inequalities: Theory, Methods, and Applications (1992)
156. C. Brezinski, Biorthogonality and Its Applications to Numerical Analysis (1992)
157. C. Swartz, An Introduction to Functional Analysis (1992)
158. S. B. Nadler, Jr., Continuum Theory: An Introduction (1992)
159. M. A. Al-Gwaiz, Theory of Distributions (1992)
160. E. Perry, Geometry: Axiomatic Developments with Problem Solving (1992)
161. E. Castillo and M. R. Ruiz-Cobo, Functional Equations and Modelling in Science and Engineering (1992)

Additional Volumes in Preparation

GEOMETRY
Axiomatic Developments with Problem Solving

Earl Perry
West Georgia College
Carrollton, Georgia

Marcel Dekker, Inc.   New York • Basel • Hong Kong

Library of Congress Cataloging-in-Publication Data

Perry, Earl.
  Geometry : axiomatic developments with problem solving / Earl Perry.
    p. cm. -- (Monographs and textbooks in pure and applied mathematics : 160)
  Includes bibliographical references and index.
  ISBN 0-8247-8727-7 (alk. paper)
  1. Geometry. I. Title. II. Series.
QA445.P46 1992
516--dc20   92-8430
                                                          CIP

This book is printed on acid-free paper.

Copyright © 1992 by MARCEL DEKKER, INC. All Rights Reserved

Neither this book nor any part may be reproduced or transmitted in any form or by any means, electronic or mechanical, including photocopying, microfilming, and recording, or by any information storage and retrieval system, without permission in writing from the publisher.

MARCEL DEKKER, INC.
270 Madison Avenue, New York, New York 10016

Current printing (last digit):
10 9 8 7 6 5 4 3 2 1

PRINTED IN THE UNITED STATES OF AMERICA

PREFACE

College geometry is a course which might be considered an endangered species. Indeed, if it were not for the fact that it is often a required course in the curriculum which leads to certification in secondary school mathematics, it might very well be extinct by now. It is truly unfortunate that geometry has been relegated to a position of relative unimportance among the courses which constitute the curriculum in mathematics, for it is an area which is both intellectually stimulating and quite utilitarian in nature.

A college course in geometry need not be simply a reworking of high-school Euclidean geometry. Neither does it need to be a study of purely non-Euclidean geometry. Unfortunately, however, in many colleges the undergraduate geometry course is often no more than a cursory review of the topics of high school geometry in which "proofs" are developed in the traditional "T" form, and tests rarely require more than the regurgitation of memorized proofs. All too frequently, courses are taught from textbooks which contain numerous unrelated, independent chapters. Since there is little or no continuity in topics, the course generally involves the memorization of some historical facts about famous geometers, the memorization of some "classical" geometry proofs developed centuries ago, a very brief and incomplete study of hyperbolic geometry topics, and possibly a "research" paper containing biographical information concerning a geometer who lived at least a hundred years ago.

Junior- and senior-level undergraduate mathematics majors, in-service school mathematics teachers and first-year graduate students in mathematics are capable of handling a good axiomatic development of geometric topics and should not be subjected to the types of courses described above. In addition, they deserve to be exposed to such supplementary geometric topics as transformational geometry, geometric constructions, and problem-solving involving geometry.

In this book, I have attempted to present materials which can be used in the teaching of a mathematically sound course in geometry which is relevant to the needs of all the groups described in the preceding paragraph. The basic material presents an orderly study of geometry, using an axiomatic approach. This approach is certainly not a unique one, nor is all the material in this book appearing in print for the first time. Indeed, much of the material has been presented by other authors, although none has presented the breadth of material which appears here. This, however, is certainly not an unusual occurrence in a mathematics book.

The content and arrangement of topics in this book resulted from a pair of courses in college geometry which I taught for twenty years. Since I was unable to locate a text which I considered to be totally satisfactory, I developed my own set of notes, which served as the basis for this book.

The book is divided into three parts: *Part I: An Introduction to Axiomatic Systems*; *Part II: An Axiomatic Development of Elementary Geometry*; and *Part III: Some Topics in Euclidean and Non-Euclidean Plane Geometry*.

In *Part I*, there are three chapters which serve as an introduction to the concept of an axiomatic system. The first contains an abbreviated study of logic and how it relates to an axiomatic system and the proving of theorems. The second chapter is used to illustrate, in a geometric setting, the techniques which are presented in Chapter 1. In this second chapter, the axioms of J. W. Young are used to develop a collection of theorems which lead to a finite geometry. The finiteness of this geometry permits the reader to study a geometric system in its entirety. The third chapter of this part contains a discussion of the properties of an axiomatic system as presented by David Hilbert in his *Foundations of Geometry*. The finite geometry of Chapter 2 provides examples for the topics in the third chapter.

The topic of study throughout *Part II* is a geometry based on the axioms which Hilbert presented in his *Foundations*. Unlike Chapter 2, in which proofs are presented for all theorems, most of the theorems in this part are presented without proof in order that the student may have an opportunity to develop his own proofs, and, hence, develop skills in studying and employing an axiomatic approach to mathematics. The material in this part is based on Hilbert's various axioms and is presented in five chapters ranging from foundations materials to materials concerning space geometry.

There are three topics presented in *Part III*. Some of the classical characteristics of Euclidean triangles and circles are examined in Chapter 9, while Chapter 10 contains an examination of the Euclidean concepts of cross ratios and harmonic sets of points and lines. Chapter 11 contains an examination of three topics from hyperbolic geometry. While the topics in *Part III* may appear to be somewhat unrelated to each other and to the materials presented in *Part II*, this is certainly not the case. Since Hilbert's works were essentially a refinement of Euclidean geometry, much of the material in *Part III* can be based on the material in *Part II*. Materials from the Appendices can be employed when *Part II* cannot be used. This results in a generally cohesive axiomatic development.

Because of my belief that a study of geometry should involve more than simply a study of axiomatic systems, I have included materials relating to three other areas of geometry—geometric transformations, geometric constructions, and geometric problem-solving. These materials appear in supplements at the end of most of the chapters. They can be omitted or used in part, since they are not related to the mainstream topics of the book except that they are topics in geometry. All of these supplementary materials are ones which I regularly included in the courses which I taught.

In my teaching of this material, I presented it in a two-quarter sequence of courses which involved approximately twenty two-hour class sessions each. While I presented the contents of *Part I* in a traditional manner, I always used a modified Moore-method approach in teaching the remaining materials. In this approach, I presented the proofs of a few theorems and allowed the students to develop and present their own proofs for the other theorems. Throughout the courses, various exercises from the Supplementary Topics were presented and assigned.

There are actually several ways that materials could be selected and presented in different formats for various courses. It certainly would be possible to omit all of *Part I* and some or all of the Supplementary Topics, particularly if the course were being presented to well-prepared mathematics majors. Another option would be to present a course composed of *Part I* and *Part II*, with or without any of the Supplementary Topics. A third option, which might very well be ideal for a course for prospective middle grades mathematics teachers, could include Chapter 1, Chapter 2, and various Supplementary Topics.

In closing, I wish to acknowledge the many persons who have provided encouragement and support for this project. My relatives and friends have been very thoughtful and understanding as well as being extremely supportive. The hundreds of West Georgia College undergraduate and graduate students who enrolled in my College Geometry I and II classes offered much encouragement and many useful suggestions. Two of my colleagues at West Georgia College have been particularly helpful. Dr. C. R. Pittman has used the materials in this book for instructing senior/graduate level classes on two occasions, and Mr. Danny Sharpe has been very unselfish with his assistance as I mastered the hardware and associated software programs I needed to produce this manuscript on a Macintosh computer. This project certainly may not have reached fruition without the encouragement and support of Ms. Maria Allegra and Mr. Walter Brownfield of Marcel Dekker, Inc. Finally, I wish to acknowledge the indirect influence of the several mathematics professors at Auburn University who introduced me to the excitement of the Moore method of instruction while I was a doctoral student in topology during the late 1960's.

Earl Perry

CONTENTS

Preface   iii

PART I: An Introduction to Axiomatic Systems   1

1 Logical Systems and Basic Laws of Reasoning   2
    1.1 Introduction   2
    1.2 The Elements of a Logical System   2
    1.3 Inductive and Deductive Reasoning   4
    1.4 Logic and Arguments   5
    1.5 Basic Forms of Logic   6
    1.6 Basic Argument Patterns   10
    Supplementary Topics
    Constructions   15
    Problems   16
    Transformation Geometry   17

2 A Simple Logical System—A Finite Geometry   23
    2.1 Introduction   23
    2.2 The Initial Undefined Terms and Axioms of the System   23
    2.3 Some Theorems of the System   25
    2.4 A Fifth Axiom and Some More Theorems   28
    2.5 A Final Axiom and Some More Theorems   38
    Supplementary Topics
    Constructions   44
    Problems   45
    Transformation Geometry   46

3 Properties of an Axiomatic System   50
    3.1 Introduction   50
    3.2 Consistency of a System   50
    3.3 Independence of Axioms   56
    3.4 Completeness of a System   59
    Supplementary Topics
    Constructions   63
    Problems   63
    Transformation Geometry   64

PART II: An Axiomatic Development of Elementary Geometry   75

4 The Foundations of Plane Geometry   76
    4.1 Introduction   76
    4.2 Existence and Incidence   76

4.3	Between—An Order Relation	78
4.4	Halflines; Partition and Separation	82
4.5	Segments; An Additional Axiom	86
4.6	Halfplanes	89
4.7	Some Additional Theorems Concerning Existence and Incidence	92
4.8	Convexity	93

Supplementary Topics
Constructions 95
Problems 95
Transformation Geometry 96

5 Triangles and Angles—Interiors and Exteriors — 102

5.1	Introduction	102
5.2	Triangles—Interiors and Exteriors	102
5.3	Angles—Interiors and Exteriors	107
5.4	Betweenness for Halflines	113

Supplementary Topics
Constructions 115
Problems 116
Transformation Geometry 117

6 Congruences and Comparisons — 124

6.1	Introduction	124
6.2	Segment Congruences	124
6.3	Angle and Triangle Congruences	126
6.4	Comparison of Segments	135
6.5	Comparison of Angles	138
6.6	Some Additional Theorems Concerning Angle Congruences and Comparisons	139

Supplementary Topics
Constructions 148
Problems 149
Transformation Geometry 149

7 Parallel and Perpendicular Lines; Additional Topics Concerning Angles, Triangles, and Segments — 159

7.1	Introduction	159
7.2	Parallel Lines	159
7.3	Interior Angles, Exterior Angles, Midpoints, and Bisectors	163
7.4	Perpendicular Lines	169
7.5	Some Additional Theorems Concerning Angles, Triangles, and Segments	173

Supplementary Topics
Constructions 175

	Problems	176
	Transformation Geometry	176
8	**The Foundations of Space Geometry**	**181**
	8.1 Introduction	181
	8.2 Existence and Incidence	181
	8.3 Halfspaces; Partition; Separation; Convexity	184
	8.4 Some Additional Theorems Concerning Existence and Incidence	186
	Supplementary Topics	
	Constructions	186
	Problems	187
	Transformation Geometry	188

PART III: Some Topics in Euclidean and Non-Euclidean Plane Geometry — **195**

9	**Some Euclidean Geometry of Triangles and Circles**	**196**
	9.1 Introduction	196
	9.2 Some Theorems Concerning Concurrences in Triangles	196
	9.3 The Theorems of Ceva and Menelaus	203
	9.4 The Simson Line, the Nine-Point Circle, and the Euler Line	215
	Supplementary Topics	
	Constructions	224
10	**Cross Ratio and Harmonic Sets**	**226**
	10.1 Introduction	226
	10.2 Cross Ratios of Points and Lines	226
	10.3 Harmonic Sets of Points and Lines	234
	10.4 Applications of Harmonic Sets	237
	Supplementary Topics	
	Constructions	245
11	**Some Topics from Hyperbolic Geometry**	**247**
	11.1 Introduction	247
	11.2 Parallels	248
	11.3 Trilaterals	277
	11.4 Quadrilaterals	292

APPENDICES	**313**
Appendix 1: Axioms and Theorems of Part II	**314**
Appendix 2: Some Plane Geometric Constructions	**323**

Appendix 3: Some Euclidean Theorems   337

Appendix 4: Some Fundamental Locus Theorems   343

Appendix 5: Some Area and Perimeter Formulas   345

Bibliography   347

Index   349

GEOMETRY

PART I

An Introduction to Axiomatic Systems

If a study of geometry is to involve more than the mere memorization of axioms and proofs of various theorems, it must begin with an examination of the concept of an axiomatic system.

Since the development of an axiomatic system is based on the ideas of logical reasoning, a study of some very basic logic is an essential preliminary to an examination of an axiomatic system and its properties. The first chapter contains an abbreviated study of basic logic forms and quantifiers. Truth tables are employed to illustrate valid and invalid forms of logical reasoning patterns.

In the second chapter, the axioms of J. W. Young are used to develop a collection of theorems which lead to a finite geometry. This collection of axioms and theorems serves two purposes: it offers illustrations of the logical reasoning patterns and techniques presented in Chapter 1, and it allows the reader to study a geometric system in its entirety.

To complete this study of axiomatic systems, a discussion of the properties of an axiomatic system, as described by David Hilbert in his *Foundations of Geometry*, is presented in the third chapter. The axioms and theorems from the finite geometry of Chapter 2 are used to illustrate these properties.

Chapter 1: Logical Systems and Basic Laws of Reasoning

1.1 Introduction

The concept and pattern for a logical system in mathematics were first presented around the year 300 B.C. by the Greek mathematician Euclid. The fact that very little of the geometry which appeared in his *Elements* originated with him detracts not at all from Euclid's genius. It is the brilliance with which he compiled and organized the works of his predecessors that has made his book one of the most influential works of any type ever produced.

In western civilization, it was probably the ancient Egyptians who first developed the basic geometric concepts relating to variously shaped figures and the relationships they have with other figures and themselves. The idea of a geometric proof as it is known in modern mathematics was totally foreign to the Egyptians, however, and their experience with reasoning was of a type known as **inductive**.

It was Euclid's Greek predecessors who first introduced the idea of a proof for a theorem. In fact, the **deductive** type of reasoning which these early mathematicians developed, and the resulting logical systems, form the basis on which all modern mathematics is built.

1.2 The Elements of a Logical System

The basic elements of any logical system are the undefined terms which are used in defining special words or expressions in the system. Since a definition gives the meaning of one word in terms of simpler words whose meanings are already known, any attempt to define every word or expression would lead to an endless regression of definitions, if it were not agreed that certain basic words would be left undefined. These basic elements, called **primitive terms** or **undefined terms**, may vary from one area of mathematics to another. Some of the undefined terms which are used in geometry are "point, line, set, all, every, any, there exists, at least one, at most one, only, one, two, ..." It should be readily apparent that these are words for which the reader may be assumed to have an understanding prior to beginning geometry.

The special terms of a logical system are explained, or defined, in **definitions** which use the undefined words and common, nontechnical words from the reader's language. Some of the definitions in geometry are for such words as "triangle, circle, parallelogram, congruent, similar, segment, convex set, ..."

In the same manner that some of the vocabulary of a logical system must be of a fundamental, undefined nature, there are some statements concerning the primitive terms which must be presumed as basic, so that other statements of the system can be deduced from them using logical reasoning. These basic statements, known as **axioms** or **postulates**, are simply assumptions about the undefined terms, and are accepted without proof.

In any logical system, the final components are called **theorems**. **Inductive reasoning** is used to develop **conjectures**—sophisticated guesses—about relationships which might exist among various defined terms and axioms of the system. **Deductive reasoning** is used in showing that these conjectures are true or false, with true conjectures being called **theorems**.

All of the components—undefined terms, definitions, axioms, and theorems—are essential for a **logical system** and are sometimes referred to as constituting an **axiomatic system**.

Exercises 1.2

1. The three basic criteria that mathematicians wish for a definition to possess are **simplicity**, **noncircularity**, and **unique characterization**.
    (a) Determine nontechnical definitions for these three terms.
    (b) How are these three criteria related to the concept of undefined terms?
    (c) Why are all three criteria important?

2. Define each of the following, and explain which of the words you use in your definition are undefined:
    (a) circle   (b) triangle

3. Symbols in mathematics often have different meanings, or definitions, in different uses. Discuss the following four uses of "=."
    (a) $2/3 = 6/9$ (b) $3x - 7 = 8$
    (c) $1 + \tan^2 x = \sec^2 x$ (d) $\angle A = \angle B$

4. In his *Elements*, Euclid offered numerous definitions. Three are:
    A **point** is that which has no part.
    A **line** is a breadthless length.
    A **straight line** is a line which lies evenly with the points on itself.

Are these good definitions, in terms of the criteria presented in problem 1?

1.3 Inductive and Deductive Reasoning

For as long as man has had a recorded history there has been evidence that people investigate various phenomena from curiosity, or sometimes necessity. Commonly, these phenomena have been related to the physical and biological aspects of their surroundings.

In ancient Egypt it was the flooding of the surrounding land by the Nile River that caused the inhabitants to begin recording the frequency and extent of the inundations in an attempt to predict their occurrences. The inhabitants of the Indian subcontinent did similar investigations regarding the monsoons. These studies were actually brought on more by necessity than by curiosity. Curiosity, however, played a major role in initial investigations regarding the behavior of the sun, moon, and other solar bodies.

Since the advent of modern medicine, physicians have conducted numerous investigations which have led to better diagnoses of various diseases. In such studies, the doctors must note all symptoms, and exclude none until it has been proved irrelevant.

All of these examples are instances where **inductive reasoning** has been employed. As a result of the examination of numerous occurrences of a particular phenomenon, an investigator is often able to predict what is more or less likely to occur. Generally, however, he will be unable to conclude, based on observations alone, that a particular result will take place each time. He is only able to use the laws of probability to determine the reliability of his prediction.

Only since the advent of logical systems has our world known and used **deductive reasoning**. This method of reasoning differs tremendously from the inductive method in that there is no investigation and recording of results. Instead, the deductive method is based entirely on the laws of logic. In fact, the expressions **logical reasoning** and **deductive reasoning** can be thought of as being synonymous. Using this method of reasoning, the investigator claims that his **premises**, which are statements he claims supply evidence or reasons leading to a particular **conclusion**, provide *absolutely conclusive* evidence. This differs from inductive reasoning, in which the investigator claims that his premises offer *some* evidence for the conclusion. As the reader has likely concluded by this point, it is **deductive reasoning** which is used in proving theorems of a logical system, although the theorems initially are just conjectures which are formulated on the basis of **inductive reasoning**.

Exercises 1.3

1. A physician has successfully used a particular treatment to cure a skin disorder in 299 consecutive cases, but in the 300th case of the same disorder, this treatment did not appear to work. Should the physician discard the treatment? What does the 300th case prove about the treatment?

2. How is inductive reasoning employed in the drafting of players for the National Football League?

3. How is inductive reasoning used by a department store's swimsuit buyer? Is this same type reasoning used by the store's financial officer when he pays the supplier for the swimsuits which are delivered to his store?

1.4 Logic and Arguments

Aristotle, one of Euclid's immediate Greek predecessors, is often credited with being one of the first persons to study and formulate simple laws of logic. There have been many others who have since contributed significantly to the science, but the basic patterns have remained pretty much the same. For the purposes of this text, only some very fundamental concepts need be studied.

To begin, **logic** has often been defined simply as the science of reasoning. This definition is, however, somewhat incomplete in that people use numerous types of reasoning—but only the deductive type is involved in logic.

A much better definition for **logic** is that it is the study of methods which can be employed to distinguish a correct argument from an incorrect one. Unfortunately, this definition lacks the characteristic of simplicity, for the word argument has not yet been defined. Hence, some preliminary definitions are required.

In the English language, one particular type of sentence is called declarative. "He is the greatest geometer who has ever lived" and "Hilbert is considered by most mathematicians to have been an outstanding geometer of the early twentieth century" would both constitute declarative sentences.

The second, however, represents what a logician would call a **statement**, or **proposition**, in that it can be assigned a **truth value** of "true" or "false." With the defining of the word **statement**, it is now possible to define an **argument** as being any group of statements in which one is claimed to follow from the others, which are considered as providing evidence for the truth of that one. The statements which supply the evidence are called **premises**, while the statement which is claimed to follow from the others is known as the **conclusion**. Unfortunately, not all deductive arguments are correct or "good."

A **valid argument** is one in which the premises and conclusion are related in such a manner that it is impossible for the premises to be true without the conclusion also being true.

Argument patterns involve combinations of simple and compound statements. Before patterns can be examined it is necessary that two classical laws of logic be stated and that some basic statements of logic be defined in terms of their truth values. The two classical laws of logic are:

The Law of Contradiction: No statement can be both true and false.

The Law of Excluded Middle: Any statement must be either true or false.

1.5 Basic Forms of Logic

In defining the basic statements of logic it is convenient to use some symbolism. Hence, the letters **P** and **Q** will be employed to represent two statements, and the letters T and F will be used to represent the truth values of "true" and "false." The five basic statements of logic which will be employed in developing the proofs in this text can now be defined in terms of their truth tables, which show all the possible assignments of truth values.

Negation		Conjunction			Disjunction		
P	not-P	**P**	**Q**	P and Q	**P**	**Q**	P or Q
T	F	T	T	T	T	T	T
F	T	T	F	F	T	F	T
		F	T	F	F	T	T
Table 1.1		F	F	F	F	F	F
		Table 1.2			**Table 1.3**		

If **P** is a statement, then **not-P** represents its **negation** or **denial** and is defined in **Table 1.1**. The statement **P and Q** is called a **conjunction** statement and is defined in **Table 1.2**, while **P or Q** is called a **disjunction** statement and is defined in **Table 1.3**. If **P, then Q**, symbolized as **P → Q**, is known as a **conditional statement**, or **implication**, and is defined in **Table 1.4**. Finally, **P if and only if Q**, symbolized as **P ↔ Q**, is known as a **biconditional statement** and is defined in **Table 1.5**.

The reader should note that the negation is referred to as a **simple statement**, while the other four are thought of as **compound statements**.

Conditional			Biconditional		
P	Q	P → Q	P	Q	P ↔ Q
T	T	T	T	T	T
T	F	F	T	F	F
F	T	T	F	T	F
F	F	T	F	F	T
Table 1.4			Table 1.5		

One of the preceding terms, **if P, then Q**, has associated with it three other statements which are sometimes encountered. These three statements are:

(1) **inverse: If not-P, then not-Q**.
(2) **converse: If Q, then P**.
(3) **contrapositive: If not-Q, then not-P**.

In attempting to construct proofs of theorems, these three forms are often considered. As you will see later, this is sometimes prudent but frequently creates difficulties when the statements are incorrectly used.

One of the fundamental ideas involved in the construction of valid arguments is that a statement may be replaced with one which is **logically equivalent**. Two statements are called **logically equivalent** when they have exactly the same truth value in every possible corresponding assignment of truth values to their components.

The replacement of a statement by one which is logically equivalent often occurs when the "denial" of a statement is used. As **Table 1.1** shows, a statement and its denial always have opposite truth values. Such statements are said to be **contradictory**. One of the most popular uses of this idea in theorem-proving is when the statement **not-(P → Q)** is replaced by **P and not-Q**. (You will be asked to show that these statements are logically equivalent in an exercise at the end of this section.)

One of the difficulties encountered when a person is attempting to replace a statement by its contradiction is that it is replaced not with the

LOGICAL SYSTEMS AND BASIC LAWS OF REASONING   7

contradictory statement but instead is replaced with a **contrary** statement. The problem here is that while **contrary** statements cannot both be true, they can both be false. Hence, a contrary for a given statement is sometimes referred to as "*a* denial" while a contradictory statement is called "*the* denial."

A discussion of basic forms of logic would not be complete without a reference to the concept of **quantifiers**, although they do not represent true logic forms like those defined with **Tables 1.1–1.5**. As the name implies, a **quantifier** is a statement which relates to the number of objects which possess a particular property. There are two types of quantifiers—**universal** and **existential**.

A **universal quantifier** is a statement that *all* the objects of a given type possess a particular property. **Table 1.6** contains some examples of frequently occurring quantifier statements. Although all the statements in this list have exactly the same meaning, sometimes it will be easier to use one particular form than another. Frequently, it is the *last* form which mathematicians use.

> *Every* line has at least three points on it.
> *All* lines have at least three points on them.
> *Each* line has at least three points on it.
> *Any* line has at least three points on it.
> If m is a line, then it has at least three points on it.

Table 1.6

An **existential quantifier** is a statement that there exists *at least one* object of a given type which possesses a particular property. **Table 1.7** contains examples of the two most frequently encountered existential quantifier statements. While these two statements have the same meaning, it is the *second* form which is most often used.

> *Some* line has three points on it.
> There exists at least one line m with three points on it.

Table 1.7

> "There exists at least one X of a particular type of object which does not possess property Y" is used for "not-(If X is a particular type of object, then it possesses property Y)."

> "If X is a particular type of object, then it does not possess property Y" is used for "not-(There exists at least one X of a particular type of object which possesses property Y)."

Table 1.8

As in the case of other statements used in theorem proving, it is sometimes desirable to use negations of quantifier statements. **Table 1.8** contains the general forms which are used in the negations of these types of statements. As you can see, an equivalent form of the negation is used, as was done in the case of **P and not-Q** being used for **not-(P → Q)**.

One of the major difficulties which students encounter when attempting to negate quantifiers is that they use a **contrary** instead of the **contradiction** for the negation of a **universal quantifier** statement. For example, if the statement "Every line has at least three points on it" is considered, its negation is "There exists at least one line which does not have at least three points on it." The statements "There exists at least one line with only one point on it" and "There exists no line with three points on it" are only *contrary* statements and not contradictory statements of the given statement.

The idea of a universal set U, a subset A of U, and the complement set A^c for A in U can be used to illustrate this concept. If $U = \{0,1,2,3,4,5,...\}$ and $A = \{3,4,5,6,...\}$, then $A^c = \{0,1,2\}$. Thus, the opposite of "at least 3," which is illustrated with set A, is "at most 2," which is illustrated with set A^c.

Exercises 1.5

1. Using one large table, determine truth values for the logic forms **P → Q, not-P → not-Q, Q → P,** and **not-Q → not-P**.
   (a) Are any of these statements logically equivalent?
   (b) Are any of these statements contradictory?
   (c) Are any of these statements contraries?

2. For each of the following statements, write the inverse, converse, and contrapositive.
   (a) **P → not-Q**   (b) **not-P → Q**
   (c) **not-Q → not-P** (d) **not-P → not-Q**

3. For each of the following statements, write the inverse, converse, and contrapositive. It may be helpful to symbolize them first, but answers should not include symbolism.
   (a) If a triangle has three congruent sides, then it is equilateral.
   (b) If two angles are right angles, then they are congruent.
   (c) If two triangles are congruent, then their corresponding angles are congruent.

4. How can the truth table you constructed in problem 1 be used to explain why some of the statements which you wrote in problem 3 are not true?

5. Why must the converse of every good definition be true? Which of the basic logic forms defined in this section is actually used in any good definition?

6. Use the truth table from problem 1 to answer these two questions:
   (a) Is the converse of every true conditional statement true?
   (b) Is the converse of every false conditional statement false?

7. The logic form **P ↔ Q**, read **P if and only if Q**, can be thought of as a conjunction of the two compound statements **P → Q** and **Q → P**.
   (a) Which of **P → Q** and **Q → P** represents the **P if Q** part of **P if and only if Q**?
   (b) Which of **P → Q** and **Q → P** represents the **P only if Q** part of **P if and only if Q**?

8. Using one large table, determine truth values for the logic forms **not-(P and Q)**, **not-(P or Q)**, **not-P and not-Q**, **not-P or not-Q**.
   (a) Are any of these statements logically equivalent?
   (b) Are any of these statements contradictory?
   (c) Are any of these statements contraries?

9. Using one large table, determine truth values for the logic forms **P → Q**, **not-(P → Q)**, **P and not-Q**.
   (a) Are any of these statements logically equivalent?
   (b) Are any of these statements contradictory?
   (c) Are any of these statements contraries?

10. For each of the following statements, write the contradictory statement and a contrary statement. It might be helpful to symbolize them first, but answers should not include symbolisms.
    (a) Yesterday was Sunday.
    (b) Dan bought a yellow truck.
    (c) The drama class was performing *Macbeth*.

11. Why are the statements "There exists at least one line with only one point on it" and "There exists no line with three points on it" only contrary statements and not contradictory statements for "Every line has at least three points on it"?

1.6 Basic Argument Patterns

Now that some basic forms of logic have been examined, it is possible for some argument patterns to be studied. While a formal definition for a valid argument pattern was given in **1.4**, there is a more usable definition. In

10 CHAPTER 1

terms of truth tables and truth values, a **valid argument** pattern is one in which there is no case in which all premises are true and the conclusion is false. In a similar manner, an **invalid argument** pattern is one in which there is at least one case in which all the premises are true and the conclusion is false.

When the above definitions are employed, an argument pattern can be considered as taking the overall form of a somewhat large **if P, then Q** statement in which the conjunction of *all* the premises of the argument pattern represents the **P** statement and the conclusion represents the **Q** statement. **Table 1.2** indicates that a conjunction is true only in the case where all its components are true, and **Table 1.4** indicates that a conditional statement is false only in the case where its **P** statement is true and its **Q** statement is false. Thus, it should be readily apparent that when an argument is being examined for validity, only those cases in which all premises are true really need to be considered.

When argument patterns are used in proving theorems, they are divided into two basic classes—**direct** and **indirect**. The examination of some skeletal forms of these types can be used for developing an understanding of how these forms can be employed in valid argument patterns.

In proving a theorem whose statement can be expressed in the form **A → Z**, the argument for the proof can take either of two forms and be a valid one. These forms are frequently referred to as the "direct" and "contrapositive" forms, although they are *both* classified as being **direct**. **Figure 1.1** and **Figure 1.2** give examples of these.

Theorem: A → Z
Proof: Suppose **A**.
    A → B
    B → C
    ⋮
    X → Y
    Y → Z
  ∴ **Z**,
    and **A → Z**.

Figure 1.1 "Direct" form of a direct argument.

Theorem: A → Z
Proof: Suppose **not-Z**.
    not- Z → B
    B → C
    ⋮
    X → Y
    Y → not -A
  ∴ **not-A**,
    and **not-Z → not-A**, which is logically equivalent to **A → Z**.

Figure 1.2 "Contrapositive" form of a direct argument.

It should be readily apparent from the forms in **Figure 1.1** and **Figure 1.2** that a kind of "transitivity" for conditional statements is frequently used in the proofs of theorems. (You will be asked to show in an exercise that "If

$A \to B$ and $B \to C$, then $A \to C$" is always a true statement.)

The fact that a direct argument of the type shown in **Figure 1.1** is valid can be easily determined with the use of a truth table, as in **Figure 1.3**.

Theorem: $A \to Z$

						premises		conclusion
Proof: Suppose A.	A	B	Z	A	$A \to B$	$B \to Z$	Z	
$A \to B$	T	T	T	T	T	T	T	*
$B \to Z$	T	T	F	T	T	F	F	
∴ Z,	T	F	T	T	F	T	T	
and $A \to Z$.	T	F	F	T	F	T	F	
	F	T	T	F	T	T	T	
	F	T	F	F	T	F	F	
	F	F	T	F	T	T	T	
	F	F	F	F	T	T	F	

Figure 1.3

In the *only* case (*) in which all the premises are true, the conclusion is also true. Hence, this a **valid argument** pattern. (You will be asked to show in an exercise that a direct argument of the type shown in **Figure 1.2** is also a valid one.)

Very frequently when students are learning how to develop valid proofs for theorems they employ some faulty logic by using two argument forms which are somewhat similar to the preceding two valid forms, but which are themselves **invalid**. Skeletal forms for these two **invalid** patterns are shown in **Figure 1.4** and **Figure 1.5**.

Theorem: $A \to Z$	**Theorem:** $A \to Z$
Proof: Suppose Z.	**Proof:** Suppose **not-A**.
$Z \to C$	**not-A** $\to$ C
$C \to A$	C $\to$ **not-Z**
∴ A,	∴ **not-Z**,
and $Z \to A$.	and **not-A** $\to$ **not-Z**.
Hence, $A \to Z$.	Hence, $A \to Z$.
Figure 1.4	**Figure 1.5**

The difficulties with these arguments occur in the concluding statements. You should recall that in problem 1 of **Exercises 1.5** you established that neither $Z \to A$ nor **not-A** $\to$ **not-Z** is logically equivalent to $A \to Z$.

While there are two basic beginnings which can lead to a valid proof for a $A \to Z$ type theorem using a **direct argument** pattern, there is only one beginning for a valid **indirect argument**. However, there are *three* satisfactory conclusions. **Figure 1.6**, **Figure 1.7**, and **Figure 1.8** give examples of these three types.

Theorem: A → Z
Proof: Suppose **A and not-Z**.
⋮
∴ **not-A**,
 and **A → Z**.

Figure 1.6

Theorem: A → Z
Proof: Suppose **A and not-Z**.
⋮
∴ **Z**,
 and **A → Z**.

Figure 1.7

Theorem: A → Z
Proof: Suppose **A and not-Z**.
⋮
∴ **not-R**,
 and **A → Z**.

Figure 1.8

Initially, you may be considerably confused about the supposition in each of these indirect forms. You should recall, however, that in problem 9 of **Exercises 1.5** you discovered that **P and not-Q** is logically equivalent to the denial of **P → Q**. Thus it can be said that in an indirect proof of an **A → Z** type of theorem the initial supposition is the denial of the statement **A → Z**. The logically equivalent form **A and not-Z** is employed because it is a more usable form than **not-(A → Z)**.

Since **A and not-Z** is assumed true, then by **Table 1.2** each of **A** and **not-Z** is, in fact, assumed to be true. In the form illustrated in **Figure 1.6**, the conclusion of **not-A** must be true for the argument to be valid. This means, however, that the initial assumption of **A and not-Z**, and thus **A**, being true must have been incorrect, by **Table 1.1**. Therefore, since **A and not-Z**, and thus **not-(A → Z)**, is false, then, **A → Z** must be true, and the theorem is proved. The form illustrated in **Figure 1.7** can be explained in a similar manner.

While the form in **Figure 1.8** involves the same initial supposition as the other two indirect forms, the conclusion appears to have little to do with component parts of the theorem. The **R** in the conclusion represents an axiom or a previously proved theorem. Since the truth value of **not-R** is always the opposite of the truth value for **R**, we would now have **R and not-R**, which, by **Table 1.2** must have a truth value of false. Thus, the initial supposition must not have been true, for a true statement cannot lead to a false statement in a valid argument pattern. As in the case of the other two forms, if **A and not–Z** is false, then **A → Z** must be true, and the theorem is proved.

To this point, all the discussion has been devoted to an explanation of valid argument patterns for theorems which have the form **A → Z**. Although most theorems can be stated in this form, a few theorems are known as **existence** theorems and *cannot* be stated in the **A → Z** form. An **existence theorem** states that a particular entity or number of entities exist, without any preliminary conditions having to be fulfilled. **Figure 1.9** illustrates an example of this type.

> **Theorem:** **K** exists.
> **Proof:** By Axiom *U* (or Theorem *V*), **A** exists.
> **A → B**
> **B → C**
> $\vdots$
> **H → J**
> **J → K**
> ∴ **K** exists.

Figure 1.9

As the skeletal form in **Figure 1.9** suggests, the proof of an existence theorem is a special type of direct argument which begins not with an initial supposition but, instead, with a known existence axiom or existence theorem. This initial statement is then followed with a sequence of **if P, then Q** statements as in the case of the forms in **Figure 1.1** and **Figure 1.2**. Again, as in the case of any valid argument, if all the premises are true, the conclusion must also be true.

In conclusion, you should realize that while this discussion may, at times, have appeared somewhat abstract, it is a necessary preliminary to the actual undertaking of theorem proving. Whatever vagueness which might have appeared should be dispelled in the next chapter, when proofs are constructed for some specific theorems. You should be prepared to refer frequently to this section as you develop skills in theorem proving.

Exercises 1.6

1. Use one large table to establish that **[A → B and B → C] → [A → C]** is always a true statement.

2. Use one large table to establish that the contrapositive form of a direct argument is, in fact, valid. Do the truth table using the following skeletal form:

Theorem: **A → Z**
Proof: Suppose **not-Z**.
not-Z → B
B → not-A
∴ **not-A**.

****

SUPPLEMENTARY TOPICS

CONSTRUCTIONS

The exercises which appear in the **CONSTRUCTIONS** section of each chapter's **SUPPLEMENTARY TOPICS** generally will be stated in terms of Euclidean geometry terminology. They are to be completed with the use of the basic constructions which are presented in **Appendix 2**. Occasionally, it may be helpful for you to refer to the list of standard Euclidean geometry theorems which are given in **Appendix 3** and/or the collection of basic Euclidean Locus theorems which are given in **Appendix 4**.

Before you proceed to the exercises, you should very carefully read the material in **Appendix 2** and be certain that you understand both the mechanics and rationale for each of the basic constructions. While the initial exercises are rather easy and are designed to allow you to develop proficiency in using the simple basic constructions, many of the later exercises will require analytical thinking and the use of several constructions, as well as theorems from **Appendix 3** and **Appendix 4**.

The exercises are to be done according to the following list of instructions, unless otherwise indicated:

(a) Use only a straightedge and compass; sketch a small circle around each point at which the compass needle point is placed on the paper.
(b) Give a brief, ordered list of commentary statements which include the numbers of the basic constructions used; do not provide formal proofs.
(c) Work problems on unlined, 8.5-inch by 11-inch paper, using only one side.

1. Using the given segments, construct a segment $\overline{GH}$ such that $m\overline{GH} = m\overline{AB} + 2(m\overline{CD} - m\overline{EF})$.

    A   B   C   D   E   F

2. Using the given angles, construct an angle ∠D such that
   $m\angle D = 2(m\angle A) + m\angle B - m\angle C$.

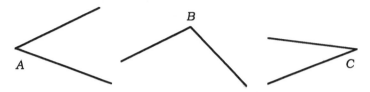

3. Construct a segment $\overline{CD}$ congruent to the given segment $\overline{HK}$. Choose any point which is not on line CD and label it as F. Construct △CDE such that F will be the midpoint of side $\overline{CE}$.

H   K

4. Construct a rhombus whose small angles measure 60° and whose sides are congruent to the given segment $\overline{AB}$.

A   B

5. Construct a rectangle such that one side is congruent to the given segment $\overline{AB}$ and a diagonal is congruent to the given segment $\overline{CD}$.

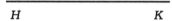

*****

PROBLEMS

The exercises which appear in the **PROBLEMS** section of each chapter's **SUPPLEMENTARY TOPICS** generally will be stated in terms of Euclidean geometry terminology. Many can be completed with the use of the materials which are presented in **Appendix 5**. While some of the exercises are relatively easy, some will require analytical thinking and the use of various standard Euclidean geometry theorems from **Appendix 3**.

The exercises are to be done according to the following list of instructions, unless otherwise indicated:
(a) Use only arithmetic, algebra, and geometry concepts in working the problems; trigonometry and calculus may *not* be used.
(b) Give a brief, ordered list of commentary statements which explain and/or justify the calculations.
(c) Work problems on 8.5-inch by 11-inch paper, using only one side.

1. Suppose a circle of radius *r* units is rolled around the outside of a circle of radius *R* units, $R > r$. If a marking instrument is attached to the smaller circle at a particular point *P*, then the pattern created by this marking instrument and the stationary large circle will be that of a stylized, petaled flower, provided *r* and *R* are related in a special way. What *is* this special way in which *r* and *R* must be related in order that there will be no "partial petals"?

✓ 2. The diagram below is composed of a square which has as its vertices four points of the larger circle. The smaller circle is tangent to each side of the square, and the two circles are concentric. If the radius of the larger circle is 3 units, determine the total area and total perimeter of the regions which are between the smaller circle and the square.

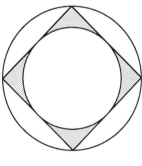

*****

TRANSFORMATION GEOMETRY

Transformation geometry is a relatively recent addition to the field of mathematics known as geometry. When it is studied in a formal, theoretical manner, transformation geometry can be seen to bring together aspects of two-space and three-space Euclidean geometry, abstract algebra, and linear algebra. Ours, however, will be a very informal study which involves definitions, examples, and exercises in a two-space setting.

T-1: Mappings and Transformations

Definition T-1: If *A* and *B* are sets of points in a plane, then a **mapping** **M** from *A* to *B* is a correspondence which associates with each point of *A* exactly one point of *B*.

Symbolism T-1: If **M** maps point P of set A to point Q of set B, then $\mathbf{M}(P) = Q$.

Definition T-2: If **M** is a mapping from a set A to a set B, then A is the **domain** of the mapping. If P is a point of A and $\mathbf{M}(P) = Q$, where Q is a point of B, then Q is the **image** of P and P is the **preimage** of Q. The subset of points of B which are images of the points of A is the **range** of the mapping.

Symbolism T-2: The image of a point P under a mapping **M** will sometimes be indicated as $\mathbf{M}(P) = P'$.

In **Figure T-1**, **M** is a mapping from set A to set B. In this mapping, $\mathbf{M}(F) = Z$, $\mathbf{M}(G) = X$, $\mathbf{M}(H) = X$. Hence, Z is the image of F, and F is the preimage of Z. Also, X is the image of each of G and H, and each of G and H is a preimage of X. Thus, A is the domain of **M**. The point Y has no preimage in A. Hence, Y is not in the range of **M**.

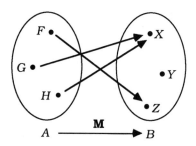

Figure T-1

Definition T-3: If **M** is a mapping from a set A to a set B in which no two distinct elements of A have the same image in B, then **M** is a **one-to-one mapping**.

In **Figure T-2**, the mapping is one-to-one, while in **Figure T-3**, the mapping is not one-to-one.

Definition T-4: If **M** is a mapping from a set A to a set B in which the range of **M** is all of B, then **M** is an **onto mapping**.

In **Figure T-4**, the mapping is onto, while in **Figure T-5**, the mapping is not onto.

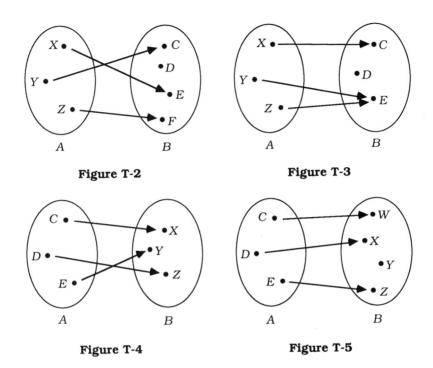

Figure T-2

Figure T-3

Figure T-4

Figure T-5

Definition T-5: A **transformation of the plane** is a one-to-one mapping of the plane onto itself.

Example T-1: Consider a mapping **M** which is defined on the Cartesian plane by $\mathbf{M}(x,y) = (x-8, y)$; that is, the point P with coordinates (x,y) has for its image the point P' with coordinates $(x-8, y)$. Thus **M** maps every point of the plane 8 units to the left. **Figure T-6** shows $\triangle ABC$ and its image $\triangle A'B'C'$. Since **M** is a one-to-one mapping of the plane onto itself, **M** is a transformation of the plane. ■

Example T-2: Consider a mapping **M** which is defined on the Cartesian plane by $\mathbf{M}(x,y) = (\sqrt{x}, y^2)$. Since $\sqrt{x}$ has no meaning when $x < 0$, then **M** does not have the entire plane for its domain. Thus, **M** is not a transformation of the plane. You should observe that **M** would also fail to be a transformation since $y^2 \geq 0$, causing **M** not to be onto the plane. (*Only one property needs to be lacking in order for a mapping to fail to be a transformation.*) ■

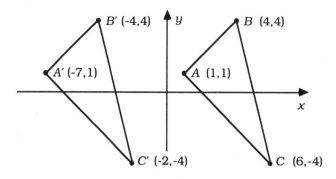

Figure T-6

Example T-3: Consider a mapping **M** which associates every point P of a plane with the "foot" P' of the perpendicular segment drawn from P to a given line L. If P is on L, then $\mathbf{M}(P) = P' = P$. This type mapping is a **projection** of the plane onto L. Clearly, **M** is not one-to-one or onto the plane, and it is not a transformation of the plane. **Figure T-7** illustrates such a mapping **M**. ■

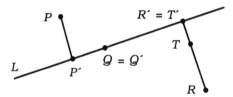

Figure T-7

Exercises T-1

1. Which of the following mappings defined on the Cartesian plane are transformations of the plane? If a mapping is not a transformation, explain why it is not.

   (a) $\mathbf{M}(x,y) = (x+3, y-4)$ (b) $\mathbf{M}(x,y) = (x-1, 3y)$

   (c) $\mathbf{M}(x,y) = (-x, -y)$ (d) $\mathbf{M}(x,y) = (x^2, y)$

   (e) $\mathbf{M}(x,y) = (x^3, y^3)$ (f) $\mathbf{M}(x,y) = (x, \ln y)$

   (g) $\mathbf{M}(x,y) = (|x|, |y|)$ (h) $\mathbf{M}(x,y) = (x, \cos y)$

2. Consider a projection **M** of the plane onto line L, and suppose C and T are two sets of points in the plane which are arranged as

shown in **Figure T-8**. (Note that neither of sets *C* and *T* includes the interior of the figure.)

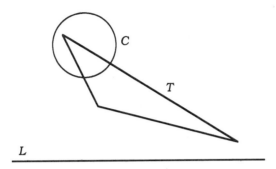

Figure T-8

(a) What will be the image of *T* under **M**? What will be the image of *C* under **M**?
(b) What will be the image of the intersection of *T* and *C*? Is it the same as the intersection of the image of *T* and the image of *C*?
(c) What will be the image of the union of *T* and *C*? Is it the same as the union of the image of *T* and the image of *C*?

3. Consider a point *O*, a line segment $\overline{AB}$, and a line *L*, as shown in **Figure T-9**. For each point *P* of $\overline{AB}$, let there correspond a point *P'* of *L* which is defined to be the intersection of ray $\overrightarrow{OP}$ and line *L*.

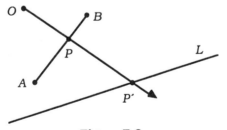

Figure T-9

(a) What would be the domain and range of this mapping?
(b) How can the preimage of any point in the range be determined?
(c) Is this mapping one-to-one?

4. Use the technique illustrated in problem 3 to devise a diagram and an informal proof for the following statement:
Any two segments contain exactly the same number of points.

5. Examine the mappings from problem 1 which you determined to be transformations of the plane, and answer the following questions:
   (a) Does a transformation preserve shape? (You may wish to consider a particular geometric object, as was done in **Example T-1**.)
   (b) Does a transformation preserve distance? That is, will the distance between P' and Q' be the same as that between P and Q, where $\mathbf{M}(P) = P'$ and $\mathbf{M}(Q) = Q'$?

6. A mapping **M** is defined for all points (x,y) of the plane by the rule $\mathbf{M}(x,y) = \begin{cases} (x, y+2), & \text{if } y \geq 0 \\ (x, y-1), & \text{if } y < 0. \end{cases}$

   (a) Determine the images of the points $(2,3)$, $(-4,5)$, $(3,-7)$, and $(0,0)$.
   (b) Determine the preimages of the points $(4,-7)$, $(-6,4)$, $(3,0)$, $(0,5)$.
   (c) Determine the domain and the range of **M**.
   (d) Is **M** one-to-one? Is **M** onto? Is **M** a transformation of the plane?
   (e) Give an example of two points P and Q and their images P' and Q' such that the distance between P' and Q' is the same as that between P and Q. Is this sufficient to show that **M** preserves distance?

7. Suppose **M** is a transformation of the plane, and determine whether each of the following statements is true or false. If a statement is false, explain why.
   (a) If P is a point, there is a unique point Q such that $\mathbf{M}(P) = Q$.
   (b) If $\mathbf{M}(P) = \mathbf{M}(Q)$, then $P = Q$.
   (c) If Q is a point, there is a unique point P such that $\mathbf{M}(P) = Q$.

8. Explain why the mapping defined by $\mathbf{M}(x,y) = (x,y)$ is usually called the **identity transformation**.

Chapter 2: A Simple Logical System—A Finite Geometry

2.1 Introduction

A very good method for developing skill in proving theorems is to begin with a logical system whose scope is limited; that is, begin with a logical system which exhibits, in miniature, the characteristics and structure of an axiomatic system. This chapter will be devoted to studying one such system, a form of which was introduced in 1909 by an American mathematician named J. W. Young. This system is called a **finite geometry** for, as you will see, it describes a system in which there are exactly nine points and twelve lines.

As the material in this chapter is presented, you should relate its form to the concepts described in Chapter 1. A good understanding of the techniques employed in this chapter is essential, if you expect to be successful in proving theorems within a logical system.

There are several approaches which can be used in developing proofs of theorems in geometry. Since axioms and theorems in a geometric setting generally pertain to entities which can be represented with a physical drawing, a **strategy** employing a collection of labeled sketches is often a very effective way for developing proofs. That technique will be employed in this chapter. As you should be certain to realize, however, pictures can sometimes be misleading. *You should be careful in your interpreting of diagrams and not draw conclusions which have no basis in axioms or theorems.*

At various places in this development there will be statements called **Notation Rules** and **Language Rules**. While these are not actually an integral part of the logical system, they will provide information concerning symbolisms and terminologies.

2.2 The Initial Undefined Terms and Axioms of the System

As you discovered in Chapter 1, an axiomatic system generally begins with some undefined terms, which are followed by axioms.

Language Rule 1: Point, line, natural numbers, and the vocabulary of sets and logic will be undefined terms.

Notation Rule 1: Points will be denoted by upper case letters, and lines will be denoted by lower case letters.

Language Rule 2: When numbers are used in the statement of axioms and theorems, they describe distinct objects.

Language Rule 3: If A is a point on a line m, the relationship between A and m can be described in several ways: **m contains A; m passes through A; A lies on m.**

Language Rule 4: If A is a point on more then one line, the lines can be described as **intersecting at A**.

Axiom 1: There is exactly one line through any two points.

Axiom 2: Every line contains at least three points.

Axiom 3: There exists at least one line.

Axiom 4: Not all points lie on the same line.

Before we look at any theorems, some discussion of these axioms will probably prove beneficial. The axioms contain mathematically correct statements, but the wording is not done in an elementary manner. A rewording of these axioms will cause them to be more understandable and, quite possibly, more usable.

Just as in the case of theorems, most axioms can be written in the form of **if P, then Q** statements. There must be, however, at least *one* existence axiom, for if there were not the axioms and theorems would describe a system which had no content. As in the case of existence theorems, described in **1.6**, an existence axiom states that an entity or number of entities of a particular type exist without any preliminary conditions having to be satisfied.

Listed below is our axiom set, rewritten in a more understandable and usable form:

Axiom 1a: If A and B are any two points, then there exists at least one line containing both A and B.

Axiom 1b: If A and B are any two points, then there exists at most one line containing both A and B.

Axiom 2: If m is a line, then there exist at least three points on m.

Axiom 3: There exists at least one line.

Axiom 4: If *m* is a line, then there exists at least one point which is not on *m*.

Language Rule 5: A line *m* which contains two points *A* and *B* can be described as having been **determined by *A* and *B***.

2.3 Some Theorems of the System

With the restatement of the axioms of the system, we are now prepared to use some of the logic discussed in Chapter 1 to prove some theorems based on the given axioms.

Theorem 1: There exist at least three points.

Strategy:

Proof: By Axiom 3, there exists a line; denote it as *m*. By Axiom 2, *m* contains at least three points.

Thus, there exist at least three points. ∎

Notation Rule 2: If points *A* and *B* determine line *m*, then the line can also be called *AB* or *BA*.

Theorem 2: There exist at least four lines.

Strategy:

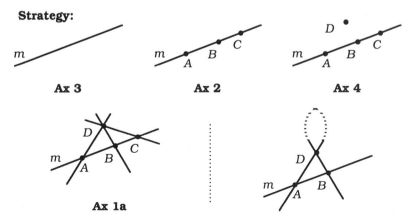

A SIMPLE LOGICAL SYSTEM—A FINITE GEOMETRY

Proof: By Axiom 3, there exists a line m. By Axiom 2, m contains at least three points; denote them as A, B, C. By Axiom 4, there exists at least one point D which is not on m.

Since A, B, C are on m and D is not on m, then D is distinct from each of A, B, C. By Axiom 1a, there exist at least one line containing A and D, one line containing B and D, and one line containing C and D; denote these lines as AD, BD, CD, respectively.

Line m is distinct from each of AD, BD, and CD because each of them contains point D, which does not lie on m.

If AD is BD, then lines m and AD would be two lines, each of which contains the two points A and B. This would be a contradiction of Axiom 1b. Thus, AD is distinct from BD.

Similarly, it can be shown that AD is distinct from CD and BD is distinct from CD.

Hence, m, AD, BD, and CD are four lines.

Thus, there exist at least four lines. ∎

You should observe that the proof of Theorem 2 is basically a direct form of an existence argument, although a major portion of the proof is based on a contradiction of an axiom. The portion of the argument in which it is established that AD is distinct from BD could have been organized more formally. In fact, this argument could have been set apart within the argument for Theorem 2. There are several accepted ways of indicating that a "little" argument is being presented within a "larger" argument. One of these ways is to label it as a **Claim**. Hence, the following **Claim** could have been placed within the proof of Theorem 2.

Claim: If AD, BD and m are lines as developed in the preceding portion of this argument (of Theorem 2), then AD is distinct from BD.

Proof: Suppose AD, BD, and m are lines as developed in the preceding portion of this argument (of Theorem 2) and AD is BD.

Since AD is BD, then, from set theory, B is a point on AD.

From earlier in the argument for this theorem, it is known that m is distinct from AD.

Hence, the two points A and B would both lie on the two lines m and AD. This is a contradiction of Axiom 1b.

Thus, the supposition is false, and the claim is true. ♦

Theorem 3: If A, B, C are three points on a line m, D is a point not on m, and AD, BD, CD are lines determined by D and A, B, C, respectively, then m, AD, BD, CD are distinct.

The statement of Theorem 2 does not indicate that Theorem 3 is true, although the proof of Theorem 3 is contained within the proof of Theorem 2. Thus, it is necessary to state the content of Theorem 3 in a separate theorem. However, Theorem 3 is presented here without proof, since, in fact, the argument presented in the preceding **Claim** would constitute a major portion of its proof.

Frequently, theorems can be stated more succinctly if the **if P, then Q** form is not used, although they are, in fact, equivalent to conditional statements. When this occurs, it is often easier to develop a proof if the theorem is restated. At other times, a restatement involving specific names of objects may cause a theorem stated as a conditional statement to be more understandable.

Theorem 4: Two lines have at most one point in common.

Restatement: If m and n are two lines, then they have at most one point in common.

Strategy:

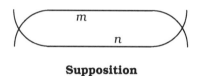

Supposition

Proof: Suppose m and n are two lines and they have at least two points in common.
This is a contradiction of Axiom 1b.
Hence, the supposition is false, and the theorem is true. ∎

Theorem 5: Not all lines pass through the same point.

Restatement: If A is a point, then there exists at least one line which does not contain A.

Strategy:

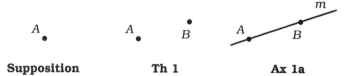

Supposition   Th 1   Ax 1a

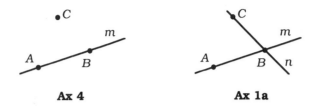

Proof: Suppose A is a point.

By Theorem 1, there exists a second point B. By Axiom 1a, A and B determine a line m.

By Axiom 4, there exists a point C which is not on m, and, hence, is distinct from B.

By Axiom 1a, B and C lie together on a line n. The lines m and n are distinct, since C is on n but not on m.

Since the two lines have the point B in common, A is distinct from B, and A is on m, then, by Theorem 4, A cannot be on n.

Hence, there exists at least one line which does not contain A. ∎

Although the **Restatement** of Theorem 5 contains the word "exist," you should observe that this is not an existence theorem like Theorem 1 and Theorem 2. In Theorem 5, the existence of a line which does not contain A is based on the preliminary condition that A is *assumed*, whereas, in Theorem 1 and Theorem 2, the existence of points and lines, respectively, is not based on any preliminary condition. As you will see, the word "exist" is frequently used in the statement of theorems which are not existence theorems.

2.4 A Fifth Axiom and Some More Theorems

In order to prove additional theorems in this system, it is necessary to use a fifth axiom suggested by Young.

Axiom 5: Through a given point not on a given line, there is exactly one line which does not intersect the given line.

As in the case of some of the other axioms presented, there are more understandable and usable forms for this axiom:

Axiom 5a: If m is a line and A is a point not on it, then there exists at least one line n which contains A but has no point in common with m.

Axiom 5b: If *m* is a line and *A* is a point not on it, then there exists at most one line *n* which contains *A* but has no point in common with *m*.

Theorem 6: There exist at least four lines through any point.

Restatement: If *A* is a point, then there exist at least four lines which contain *A*.

Strategy:

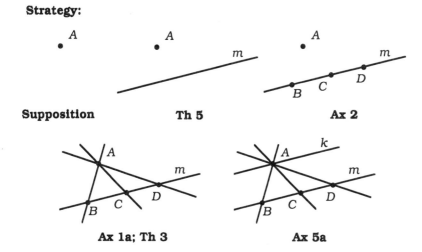

Proof: Suppose *A* is a point.

By Theorem 5, there exists a line *m* which does not contain *A*. By Axiom 2, *m* contains at least three points; denote them as *B*, *C*, *D*.

Since *B*, *C*, *D* are on *m* and *A* is not on *m*, then *A* is distinct from each of *B*, *C*, *D*. Thus, by Axiom 1a, *A* and *B* determine line *AB*, *A* and *C* determine line *AC*, and *A* and *D* determine line *AD*. By Theorem 3, the lines *AB*, *AC*, and *AD* are distinct.

Since *A* is not on *m*, then, by Axiom 5a, there exists a line *k* through *A* which has no point in common with *m*. Line *k* is distinct from each of *AB*, *AC*, *AD* since each of these lines contains a point of *m*.

Hence, there exist at least four lines which contain *A*. ∎

Definition 1: Two lines are called **parallel lines** if they have no point in common. The members of a collection of more than two lines are called parallel if no two of them have a point in common.

Theorem 7: There exist at least two lines parallel to a given line.

Restatement: If m is a line, then there exist at least two lines parallel to m.

Strategy:

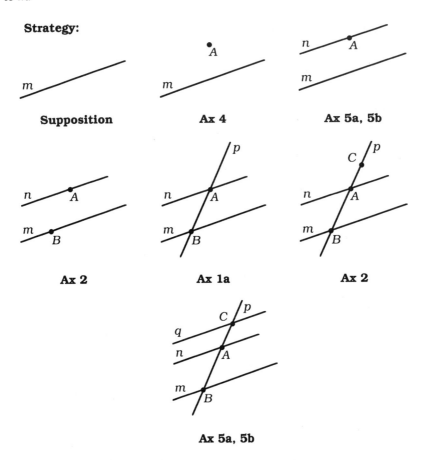

Proof: Suppose m is a line.

By Axiom 4, there exists a point A which is not on m.

By Axiom 5a and Axiom 5b, there exists exactly one line n which contains A but which has no point in common with m. Thus, by Definition 1, n is parallel to m.

By Axiom 2, there exists a point B on m. Since B is on m and A is not on m, then A and B are distinct. Thus, by Axiom 1a, there exists a line p which contains A and B. Since A is on p and A is not on m, then p and m are distinct.

By Axiom 2, p contains a third point C. If C is on m, then the two points B and C would both lie on the two lines m and p. This would contradict Axiom 1b. Hence C is not on m.

30   CHAPTER 2

By Axiom 5a and Axiom 5b, there exists exactly one line q which contains C but which has no point in common with m. Thus, by Definition 1, q is parallel to m.

Since p shares the point B with m, and each of n and q has no point in common with m, then p is distinct from each of n and q.

If n is q, then both A and C would lie on n. Thus, the two points A and C would lie on the two lines n and p. This would contradict Axiom 1b.

Hence, n and q are distinct lines, and, thus, there exist at least two lines parallel to m. ∎

Theorem 8: If a line, distinct from two parallel lines, intersects one of them, then it intersects the other.

Restatement: If m and n are two parallel lines and q is a third line which has a point in common with m, then q also has a point in common with n.

Strategy:

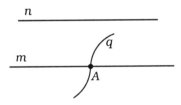

Supposition

Proof: Suppose m, n, q are three lines, m and n are parallel, q intersects m at a point A, and q does not intersect n.

Since q does not intersect n, then, by Definition 1, q and n are parallel. Thus, m and q are two lines which have no point in common with n.

Thus, m and q are two lines which contain the point A, which is not on n. This is a contradiction of Axiom 5b.

Hence, the supposition must be false, and the theorem is true. ∎

Theorem 9: Two lines parallel to the same third line are parallel to each other.

Restatement: If m, n, q are three lines such that n is parallel to m and q is parallel to m, then n and q are parallel.

Strategy:

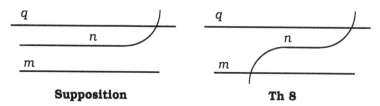

| Supposition | Th 8 |

Proof: Suppose m, n, q are three lines such that n is parallel to m, q is parallel to m, and n is not parallel to q.

Since n is not parallel to q, then, by Definition 1, n intersects q. By Theorem 8, n would also intersect m, since m is parallel to q.

Thus, n is not parallel to m, since it intersects m.

Hence, the supposition is false, and the theorem is true. ■

In the nine theorems which have been examined to this point in our study of a finite system, the only number quantities which have occurred have been one, two, three, and four. At several times, however, the theorems have contained the phrases "at least one," "at least two," "at least three," and "at least four." These phrases certainly imply that there may be more than one, two, three, or four objects, as the particular case may be.

The next three theorems differ from the previous nine in that they contain the number quantities n and $n-1$. The use of these number quantities certainly alludes to the possibility that there may be far more objects than have been shown to exist in the first nine theorems. As you will see, the proving of theorems of this type often involves more complex statements than those employed in proving Theorems 1–9.

Theorem 10: There exist exactly $n-1$ lines parallel to a given line containing exactly n points.

Restatement: If m is a line containing exactly n points, then there exist exactly $n-1$ lines parallel to m.

Strategy:

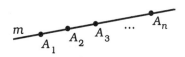

Supposition

Proof: Suppose m is a line containing exactly n points; denote them as $A_1, A_2, ..., A_n$.

Claim: There exist at least $n-1$ lines parallel to m.

Strategy:

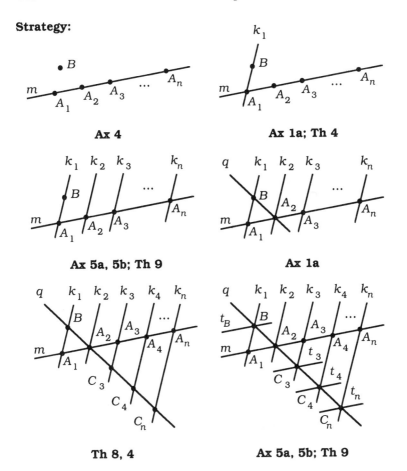

Proof: By Axiom 4, there exists a point B which is not on m, and, hence, is distinct from each of $A_1, A_2, ..., A_n$. By Axiom 1a, A_1 and B determine a line k_1. The lines k_1 and m are distinct, since B is on k_1 and B is not on m.

None of $A_2, A_3, ..., A_n$ is on k_1, by Theorem 4, since k_1 and m are distinct and A_1 is on k_1. By Axiom 5a and Axiom 5b, each of $A_2, A_3, ..., A_n$ lies on exactly one line which has no point in common with k_1; denote these lines as $k_2, k_3, ..., k_n$, respectively.

A SIMPLE LOGICAL SYSTEM—A FINITE GEOMETRY   33

By Definition 1, each of $k_2, k_3, ..., k_n$ is parallel to k_1. Hence, each pair of lines in the collection $k_2, k_3, ..., k_n$ will be parallel, by Theorem 9. Thus, again by Definition 1, the lines $k_1, k_2, ..., k_n$ are all parallel.

Since B is distinct from A_2, by Axiom 1a there exists a line q which contains them. This line q intersects k_1 at B and k_2 at A_2, and, hence, by Theorem 8, it intersects each of $k_3, k_4, ..., k_n$; denote these points as $C_3, C_4, ..., C_n$, respectively. The points $B, A_2, C_3, ..., C_n$ are distinct, since they lie on parallel lines.

The line q is distinct from m, since q contains B, which is not on m. None of the points $C_3, C_4, ..., C_n$ can lie on m, by Theorem 4, since A_2 is on both q and m and each of $C_3, C_4, ..., C_n$ is distinct from A_2.

By Axiom 5a and Axiom 5b, through each of $B, C_3, C_4, ..., C_n$, there passes exactly one line which has no point in common with m; denote these lines as $t_B, t_3, t_4, ..., t_n$, respectively. By Definition 1, each line in this collection is parallel to m.

By Theorem 9, each pair of lines in the collection $t_B, t_3, t_4, ..., t_n$ will be parallel. Thus, by Definition 1, the lines $t_B, t_3, t_4, ..., t_n$ are parallel, and, hence, distinct. Therefore, there exist at least $n-1$ lines parallel to m. ◆

Claim: There exist at most $n-1$ lines parallel to m.

Strategy:

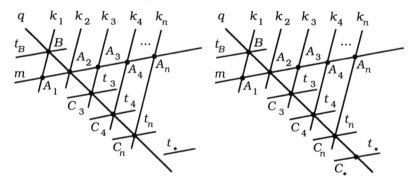

Supposition   **Th 9, 8**

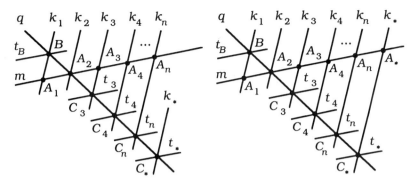

Th 4; Ax 5a; Th 9　　　　　　　**Th 8**

Proof: Suppose there exists an nth line parallel to m; denote it as t_*.

By Theorem 9, t_* will be parallel to each of t_B, t_3, t_4, ..., t_n. Since q intersects m, then, by Theorem 8, q intersects t_* in a point C_*. By Definition 1, C_* is distinct from each of B, A_2, C_3, C_4, ..., C_n. Thus, C_* will be an $(n+1)^{th}$ point on q.

Point C_* is not on k_1, by Theorem 4, since C_* and B are distinct and k_1 and q already have B in common.

By Axiom 5a and Definition 1, there exists a line k_* which passes through C_* and is parallel to k_1. By Theorem 9, k_* will be parallel to each of k_2, k_3, ..., k_n. Hence, none of A_1, A_2, ..., A_n is on k_*, by Definition 1.

Since m intersects k_1, then, by Theorem 8, m intersects k_* at a point A_*. This would cause m to contain $n+1$ points, since none of the n points A_1, A_2, ..., A_n is on k_*. This contradicts the initial supposition that m contains exactly n points.

Thus, the supposition that there exists an nth line parallel to m is false, and, therefore, there are at most $n-1$ lines parallel to m. ♦

Since there are at least $n-1$ lines parallel to m and at most $n-1$ lines parallel to m, there are exactly $n-1$ lines parallel to m. ∎

Theorem 11: There exist exactly n points on any line intersecting a given line containing exactly n points.

Restatement: If m is a line containing exactly n points and k is a line intersecting m, then k contains exactly n points.

Strategy:

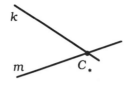

Supposition

Proof: Suppose m is a line containing exactly n points and k is a line intersecting m; denote this point of intersection as C_*.

Claim: Line k contains at least n points.

Strategy:

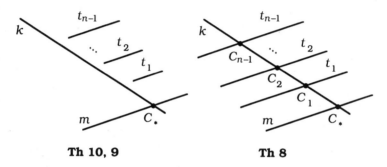

Th 10, 9   Th 8

Proof: By Theorem 10, there exist exactly $n-1$ lines parallel to m; denote these lines as $t_1, t_2, \ldots, t_{n-1}$. Each pair in this collection will be parallel, by Theorem 9. Thus, by Definition 1, the lines $m, t_1, t_2, \ldots, t_{n-1}$ are all parallel.

The line k is distinct from each of $t_1, t_2, \ldots, t_{n-1}$, since k contains a point of m, which has no point in common with any of the lines in this collection.

The line k intersects each of $t_1, t_2, \ldots, t_{n-1}$, by Theorem 8, since k intersects m. Denote these points as $C_1, C_2, \ldots, C_{n-1}$, respectively. Since $t_1, t_2, \ldots, t_{n-1}, m$ are all parallel, then the points $C_1, C_2, \ldots, C_{n-1}, C_*$ are distinct, by Definition 1.

Thus, k contains at least n points. ♦

Claim: Line k contains at most n points.

Strategy:

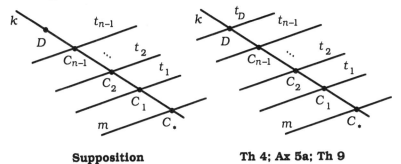

Supposition   Th 4; Ax 5a; Th 9

Proof: Suppose k contains an $(n+1)^{th}$ point D.

Since k and m are distinct and share the point C_*, then, by Theorem 4, D is not on m. Similarly, D is not on any of $t_1, t_2, ..., t_{n-1}$ since k is distinct from each of these lines and shares with them the points $C_1, C_2, ..., C_{n-1}$, respectively.

By Axiom 5a, there exists a line t_D which contains D and is parallel to m. Since t_D contains D, which is not on any of $t_1, t_2, ..., t_{n-1}$, then t_D is distinct from each of $t_1, t_2, ..., t_{n-1}$. By Theorem 9, t_D is parallel to each of $t_1, t_2, ..., t_{n-1}$.

Thus, there exist at least n lines parallel to m, contradicting Theorem 10. Hence, the supposition is false, and there exist at most n points on k. ♦

Since there are at least n points on k and at most n points on k, there exist exactly n points on k. ∎

Theorem 12: There exist exactly n points on any line parallel to a given line containing exactly n points.

Restatement: If m is a line containing exactly n points and k is a line parallel to m, then k contains exactly n points.

Strategy:

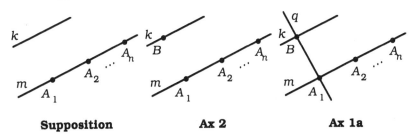

Supposition   Ax 2   Ax 1a

A SIMPLE LOGICAL SYSTEM—A FINITE GEOMETRY

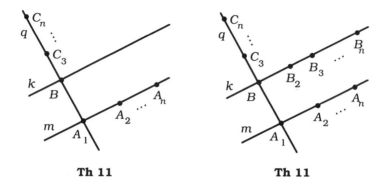

Th 11   Th 11

Proof: Suppose m is a line containing exactly n points and k is a line parallel to m. Denote the points on m as A_1, A_2, ..., A_n.

By Axiom 2, there exists a point B on k. Since m and k are parallel, then, by Definition 1, B and A_1 are distinct.

By Axiom 1a, B and A_1 lie together on a line q. Since B is on k and A_1 is on m, then q intersects each of k and m.

Line q contains exactly n points, by Theorem 11, since it intersects m, which contains exactly n points.

Hence, by Theorem 11, k contains exactly n points, since it intersects q, which contains exactly n points. ∎

2.5 A Final Axiom and Some More Theorems

In the introduction to this chapter, it was stated that the logical system to be studied was based on an axiom set presented originally by J. W. Young. The observation was made that this set of axioms applies to a finite geometry in which there are exactly nine points and twelve lines. To this point in our study we have established far less than these numbers of points and lines. In fact, we have shown only the existence of at least three points, in Theorem 1, and at least four lines, in Theorem 2.

Although the statements in Theorems 10–12 allude to the possibility of there being more points and lines in this system, the number represented by "n" is never specified. In order to give a value to this number n, it is necessary to use the last of the axioms suggested by Young.

Axiom 6: Every line contains at most three points.

As in the case of some of the axioms presented initially in this study, a more understandable and usable form of this axiom can be stated:

Axiom 6: If m is a line, then there exist at most three points on m.

With this axiom, we can now prove that our system is, indeed, finite in nature.

Theorem 13: There exist exactly nine points.

Strategy:

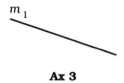

Ax 3

Proof: By Axiom 3, there exists a line m_1.

Claim: There exist at least nine points.

Strategy:

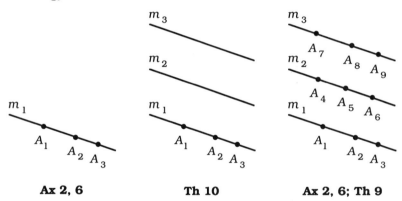

Ax 2, 6   **Th 10**   **Ax 2, 6; Th 9**

Proof: By Axiom 2 and Axiom 6, there exist exactly three points on m_1; denote these points as A_1, A_2, A_3.

By Theorem 10, with $n = 3$, there exist exactly two lines parallel to m_1; denote these lines as m_2 and m_3. By Axiom 2 and Axiom 6, each of m_2 and m_3 contains exactly three points; denote these points as A_4, A_5, A_6 and A_7, A_8, A_9, respectively.

The lines m_2 and m_3 are parallel, by Theorem 9. Hence, by Definition 1, no two of the parallel lines m_1, m_2, m_3 have a point in common.

Thus, the points A_1, A_2, ..., A_9 are distinct. Therefore, there exist at least nine points. ◆

Claim: There exist at most nine points.

Strategy:

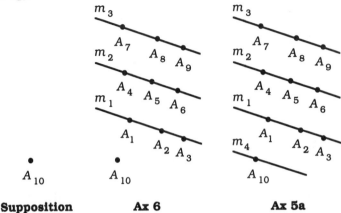

| Supposition | Ax 6 | Ax 5a |

Proof: Suppose there exists a tenth point; denote it as A_{10}.

By Axiom 6, A_{10} cannot be on any of the lines m_1, m_2, m_3.

Since A_{10} is not on m_1, then, by Axiom 5a, there exists a line m_4 containing A_{10} and having no point in common with m_1. By Definition 1, m_4 is parallel to m_1.

The line m_4 is distinct from each of m_2 and m_3, since A_{10} is on neither of them.

Thus, m_2, m_3, m_4 are three lines which are parallel to the line m_1, which contains exactly three points. This is a contradiction of Theorem 10.

Hence, the supposition is false, and there exist at most nine points. ◆

Since there exist at least nine points and there exist at most nine points, there exist exactly nine points. ■

Theorem 14: There exist exactly twelve lines.

Strategy:

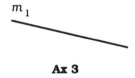

Ax 3

Proof: By Axiom 3, there exists a line m_1.

Claim: There exist at least twelve lines.

Strategy:

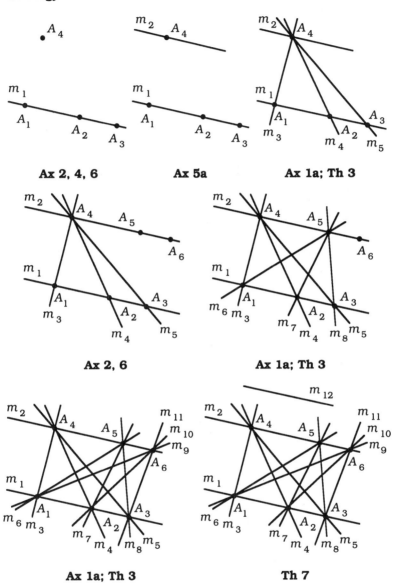

Proof: By Axiom 2 and Axiom 6, there exist exactly three points on m_1; denote these points as A_1, A_2, A_3. By Axiom 4, there exist a point A_4

A SIMPLE LOGICAL SYSTEM—A FINITE GEOMETRY   41

which is not on m_1. Since A_4 is not on m_1, then A_4 is distinct from each of A_1, A_2, A_3.

By Axiom 5a, there exists a line m_2 containing A_4 and having no point in common with m_1.

By Axiom 1a, A_4 lies on a line with each of A_1, A_2, A_3; denote these lines as m_3, m_4, m_5, respectively. By Theorem 3, m_3, m_4, m_5 are distinct. Since each of m_3, m_4, m_5 contains A_4, which is not on m_1, each of these lines is distinct from m_1.

Line m_2 is parallel to m_1, by Definition 1, and distinct from each of m_3, m_4, m_5, since each shares a point with m_1.

By Axiom 2 and Axiom 6, m_2 contains exactly two other points; denote these points as A_5 and A_6.

Since A_5 is not on m_1, it is distinct from each of A_1, A_2, A_3. By Axiom 1a, A_5 lies on a line with each of these three points; denote these lines as m_6, m_7, m_8, respectively. By Theorem 3, these three lines are distinct.

The lines m_6, m_7, m_8 are distinct from m_1, since each contains point A_5, which is not on m_1. Also, each of these three lines is distinct from m_2, since each contains a points of m_1, which is parallel to m_2.

The lines m_3 and m_6 are distinct, for if they were not then the line m_3, which is distinct from m_2, would have the two points A_4 and A_5 in common with m_2. This would contradict Theorem 4. Analogously, m_3 is distinct from each of m_7 and m_8, m_4 is distinct from each of m_6, m_7, m_8, and m_5 is distinct from each of m_6, m_7, m_8.

In a manner similar to the preceding, it can be shown that A_6 lies on a line with each of A_1, A_2, A_3. These lines, which can be denoted as m_9, m_{10}, m_{11}, can be proved distinct from each other and from each of m_1, ..., m_8 in an analogous manner.

Of these distinct lines m_1, ..., m_{11}, all but m_2 have a point in common with m_1. Hence, by Theorem 7, there exists a second line parallel to m_1 and distinct from m_2; denote it as m_{12}. By Definition 1, m_{12} is distinct from m_1. Since each of the lines m_3, ..., m_{11} contains a point of m_1, which is parallel to m_{12}, then each of m_3, ..., m_{11} is distinct from m_{12}.

Thus, there exist at least twelve lines. ◆

Claim: There exist at most twelve lines.

Strategy: See **Supposition** diagram at the top of page 43.

Proof: Suppose there exists a thirteenth line; denote it as m_{13}.

Since m_1 contains exactly three points, and each of m_2 and m_{12} is parallel to m_1, then, by Theorem 10, m_{13} cannot be parallel to m_1. By

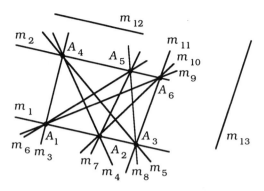

Supposition

Definition 1, m_{13} must have a point in common with m_1, and, hence, by Theorem 8, m_{13} must have a point in common with m_2.

Each of m_1 and m_2 contains exactly three points, namely A_1, A_2, A_3 and A_4, A_5, A_6, respectively. These points already lie together in pairs on the nine lines $m_3, ..., m_{11}$.

If m_{13} is to intersect each of m_1 and m_2, which contain no points other than A_1, A_2, A_3, and A_4, A_5, A_6, respectively, then m_{13} would contain the same pair of points as one of $m_3, ..., m_{11}$. This would contradict Theorem 4.

Thus, the supposition that there exists a thirteenth line is false.

Hence, there exist at most twelve lines. ◆

Since there exist at least twelve lines and at most twelve lines, there exist exactly twelve lines. ∎

Exercises 2.5

In 1892, an Italian mathematician named G. Fano introduced the following set of axioms:

Axiom 1: There is exactly one line through any two points.
Axiom 2: Every line contains at least three points.
Axiom 3: There exists at least one line.
Axiom 4: Not all points lie on the same line.

Axiom 5: Any two lines have a point in common.
Axiom 6: Every line contains at most three points.

1. Reexamine the proofs of the theorems in this chapter and determine which are also true using Fano's axioms and, hence, can be used to prove other theorems in Fano's axiomatic system.

2. Use **Axioms 1–4** of Fano's set to prove:
   Theorem: There exist at least three lines through any point.

3. Use **Axioms 1–5** of Fano's set to prove:
   (a) **Theorem:** If there exists a line containing exactly n points, then every line contains exactly n points.
   (b) **Theorem:** If there exists a line containing exactly n points, then every point has exactly n lines through it.

4. Use **Axioms 1–6** of Fano's set to prove:
   (a) **Theorem:** There exist exactly seven points.
   (b) **Theorem:** There exist exactly seven lines.

****

SUPPLEMENTARY TOPICS

CONSTRUCTIONS

6. Construct an isosceles triangle whose congruent angles measure 45° and whose hypotenuse is congruent to the given segment $\overline{AB}$.

   A   B

7. Construct a 30°–60°–90° right triangle in which the side opposite the 30° angle is congruent to the given segment $\overline{AB}$.

   A   B

8. Construct a circle and a line tangent to this circle. Using an initial segment measuring 1 inch, determine all points which lie on the tangent line and are 2 inches from the circle.

9. Construct a segment $\overline{CD}$ congruent to the given segment $\overline{AB}$. Determine the location of the point E on $\overline{CD}$ such that $m\overline{CE} = 0.4(m\overline{CD})$.

A ——————————————————————— B

*****

PROBLEMS

3. In the diagram below, two semicircles have sides of the square as diameters. If a side of the square has measure 6 units, what are the area and perimeter of the shaded region?

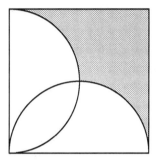

4. When a particular electric fan is operated on "low," its blades turn at 50 revolutions per minute. The distance from the center of the fan to the tip of each blade is 6 inches. If a speck of paint is on the tip of one of the blades, what is the velocity, in feet per second, of the speck when the fan is running on "low"?

5. In the following diagram, the two circles are concentric, and the four circular arcs tangent to the smaller circle are based on the same radius as that used for the larger circle. If the radius of the larger circle is 3 units, determine the total area and total perimeter of the shaded regions.

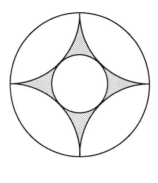

*****

TRANSFORMATION GEOMETRY

T-2: Isometries

Definition T-6: A transformation of the plane which preserves distance is an **isometry**.

Example T-4: The mapping $\mathbf{M}(x,y) = (x-8,y)$ is a transformation of the plane. If $P_1 = (x_1,y_1)$ and $P_2 = (x_2,y_2)$, then $\mathbf{M}(P_1) = (x_1-8,y_1)$ and $\mathbf{M}(P_2) = (x_2-8,y_2)$.

The distance between P_1 and P_2 is $\sqrt{(x_2-x_1)^2 + (y_2-y_1)^2}$ and the distance between $\mathbf{M}(P_1)$ and $\mathbf{M}(P_2)$ is

$$\sqrt{[(x_2-8)-(x_1-8)]^2 + (y_2-y_1)^2} = \sqrt{(x_2-x_1)^2 + (y_2-y_1)^2}.$$

Since these distances are the same, **M** is also an isometry. ∎

Because of the informality of this study, theorems will be stated without proof.

Theorem T-1: An isometry preserves collinearity.

Example T-5: Consider the isometry $\mathbf{M}(x,y) = (x,y-2)$, and suppose $A(x_1,y_1)$, $B(x_2,y_2)$, $C(x_3,y_3)$ are three collinear points with B between A and C. Then

$$\sqrt{(x_2-x_1)^2+(y_2-y_1)^2}+\sqrt{(x_3-x_2)^2+(y_3-y_2)^2} = \sqrt{(x_3-x_1)^2+(y_3-y_1)^2}.$$

Since $\mathbf{M}(A)=(x_1,y_1-2)$, $\mathbf{M}(B)=(x_2,y_2-2)$, $\mathbf{M}(C)=(x_3,y_3-2)$, the distances between $\mathbf{M}(A)$ and $\mathbf{M}(B)$, $\mathbf{M}(B)$ and $\mathbf{M}(C)$, $\mathbf{M}(A)$ and $\mathbf{M}(C)$ are, respectively,

$$\sqrt{(x_2-x_1)^2+[(y_2-2)-(y_1-2)]^2}, \sqrt{(x_3-x_2)^2+[(y_3-2)-(y_2-2)]^2},$$

$$\sqrt{(x_3-x_1)^2+[(y_3-2)-(y_1-2)]^2}.$$

Since

$$\sqrt{(x_2-x_1)^2+[(y_2-2)-(y_1-2)]^2}+\sqrt{(x_3-x_2)^2+[(y_3-2)-(y_2-2)]^2}$$

$$=\sqrt{(x_2-x_1)^2+(y_2-y_1)^2}+\sqrt{(x_3-x_2)^2+(y_3-y_2)^2}$$

and $\sqrt{(x_3-x_1)^2+[(y_3-2)-(y_1-2)]^2}=\sqrt{(x_3-x_1)^2+(y_3-y_1)^2}$,

then $\sqrt{(x_2-x_1)^2+[(y_2-2)-(y_1-2)]^2}+\sqrt{(x_3-x_2)^2+[(y_3-2)-(y_2-2)]^2}$

$$=\sqrt{(x_3-x_1)^2+[(y_3-2)-(y_1-2)]^2}.$$

Hence, $\mathbf{M}(B)$ is between $\mathbf{M}(A)$ and $\mathbf{M}(C)$, and $\mathbf{M}$ preserves collinearity. ∎

Theorem T-2: An isometry maps lines to lines, segments to segments, rays to rays, and angles to angles.

Theorem T-3: An isometry preserves parallelism.

Example T-6: Consider the transformation $\mathbf{M}(x,y)=(x-1,y-2)$, and suppose L_1 and L_2 are two parallel lines. If $A(x_1,y_1)$ and $B(x_2,y_2)$ are points on L_1, and $C(x_3,y_3)$ and $D(x_4,y_4)$ are points on L_2, then

$$\frac{y_2-y_1}{x_2-x_1}=\frac{y_4-y_3}{x_4-x_3},$$

since L_1 and L_2 are parallel.

Now $\mathbf{M}(A) = (x_1-1, y_1-2)$, $\mathbf{M}(B) = (x_2-1, y_2-2)$, $\mathbf{M}(C) = (x_3-1, y_3-2)$, and $\mathbf{M}(D) = (x_4-1, y_4-2)$, where $\mathbf{M}(A)$ and $\mathbf{M}(B)$ are on $\mathbf{M}(L_1) = L_1'$ and $\mathbf{M}(C)$ and $\mathbf{M}(D)$ are on $\mathbf{M}(L_2) = L_2'$.

The slope of L_1' is

$$\frac{(y_2-2)-(y_1-2)}{(x_2-1)-(x_1-1)} = \frac{y_2-y_1}{x_2-x_1},$$

while the slope of L_2' is

$$\frac{(y_4-2)-(y_3-2)}{(x_4-1)-(x_3-1)} = \frac{y_4-y_3}{x_4-x_3}.$$

Since

$$\frac{y_2-y_1}{x_2-x_1} = \frac{y_4-y_3}{x_4-x_3},$$

then L_1' and L_2' are parallel. Thus $\mathbf{M}$, preserves parallelism. ∎

Theorem T-4: An isometry preserves angle measure.

Example T-7: Consider the isometry $\mathbf{M}(x,y) = (x-5, y+3)$, $\angle ABC$, and $\mathbf{M}(\angle ABC) = \angle A'B'C'$, where $\mathbf{M}(A) = A'$, $\mathbf{M}(B) = B'$, $\mathbf{M}(C) = C'$ for $A(x_1, y_1)$, $B(x_2, y_2)$, $C(x_3, y_3)$.

Thus,

$$\mathbf{M}(A) = (x_1-5, y_1+3), \mathbf{M}(B) = (x_2-5, y_2+3), \mathbf{M}(C) = (x_3-5, y_3+3).$$

The distances between A and B, B and C, C and A are, respectively,

$$\sqrt{(x_2-x_1)^2 + (y_2-y_1)^2}, \sqrt{(x_3-x_2)^2 + (y_3-y_2)^2}, \sqrt{(x_3-x_1)^2 + (y_3-y_1)^2},$$

while those between A' and B', B' and C', C' and A' are, respectively,

$$\sqrt{[(x_2-5)-(x_1-5)]^2 + [(y_2+3)-(y_1+3)]^2},$$

$$\sqrt{[(x_3-5)-(x_2-5)]^2 + [(y_3+3)-(y_2+3)]^2},$$

$$\sqrt{[(x_3-5)-(x_1-5)]^2 + [(y_3+3)-(y_1+3)]^2}.$$

Since these last three expressions simplify to

$$\sqrt{(x_2-x_1)^2+(y_2-y_1)^2},\ \sqrt{(x_3-x_2)^2+(y_3-y_2)^2},\ \sqrt{(x_3-x_1)^2+(y_3-y_1)^2},$$

then $\triangle A'B'C'$ would be congruent to $\triangle ABC$, so corresponding parts would be congruent.

Hence, the measure of $\angle A'B'C'$ would be the same as the measure of $\angle ABC$. Thus, **M** preserves angle measure. ∎

Exercises T-2

1. Which of the following transformations of the plane are isometries?
   (a) $\mathbf{M}(x,y) = (x+1, y-5)$ (b) $\mathbf{M}(x,y) = (x,y)$
   (c) $\mathbf{M}(x,y) = (x^3, y)$ (d) $\mathbf{M}(x,y) = (-x, y+1)$
   (e) $\mathbf{M}(x,y) = (3x, 3y)$ (f) $\mathbf{M}(x,y) = (½x, -⅓y)$

2. Does an isometry preserve perpendicularity? Why?

3. Does an isometry map the midpoint of a segment to the midpoint of the image of the segment? Why?

4. Consider the transformation $\mathbf{M}(x,y) = (-2x+5, 2y-3)$.
   (a) Does **M** preserve collinearity?
   (b) Does **M** preserve parallelism?
   (c) Does **M** preserve angle measure?
   (d) Is **M** an isometry?

5. Develop an informal proof for Theorem T-1.

6. Develop an informal proof for Theorem T-2.

7. Develop an informal proof for Theorem T-3.

8. Develop an informal proof for Theorem T-4.

Chapter 3: Properties of an Axiomatic System

3.1 Introduction

To this point in your study, you have examined two major concepts relative to an axiomatic system. In Chapter 1, you studied the fundamental ideas of logic which are associated with the development of such a system. This study was followed immediately, in Chapter 2, by an in-depth examination of a somewhat concise system which described a finite geometry.

In order to conclude an analysis of axiomatic systems, you need to examine three basic properties of axiomatic systems. This chapter will be devoted to the study of these properties: **consistency** of an axiomatic system; **independence** of axioms within a set of axioms; and **completeness** of an axiomatic system. As you will observe, frequent references will be made to ideas presented in Chapter 1 and to the axioms and theorems studied in Chapter 2.

3.2 Consistency of a System

Of the three properties which are associated with an axiomatic system, only the property of consistency is *absolutely essential*. The necessity of consistency should be readily apparent from the following definition:

Definition 1: An axiomatic system **S** is **consistent** if it contains no contradictory statements.

To put this idea less formally, this definition states that in order for a system to be consistent it cannot contain any two axioms, any axiom and theorem, or any two theorems of the forms **P** and **not-P**.

As you should realize, there is a fundamental difficulty which arises immediately: *How can you be certain that all the axioms and all the theorems for a particular system have been examined?* Obviously, the answer is that you cannot be certain. A closely related difficulty may arise when the theorems developed in a particular system become so numerous and complicated that a pair of contradictory ones might not be detected. As you can easily imagine, both of these difficulties are more likely to arise when the system being considered relates to an infinite number of objects. For these reasons, it is essentially impossible to establish absolute consistency for a system. There is, however, a test which can be used to show **relative**

consistency. Before this test is considered, some more definitions are needed.

Definition 2: An **interpretation** of an axiomatic system **S** is the assignment of meanings to the undefined technical terms of **S** in such a manner that all the axioms become true or false statements.

Definition 3: An axiomatic system **S** is **satisfiable** if there exists an interpretation which causes all its axioms to be true.

Definition 4: An interpretation of an axiomatic system **S** which causes it to be satisfied is a **model** for **S**.

Test for Consistency: An axiomatic system **S** is **relatively consistent** if and only if there exists at least one model for it.

As the preceding definitions and test statement indicate, the key to establishing that an axiomatic system is relatively consistent is the discovery of an interpretation which makes all the axioms in the system true. Since the proof of each theorem in a system can be considered as taking the overall form of a somewhat large **if P, then Q** statement in which the conjunction of all the premises of the argument pattern represents the **P** statement and the conclusion represents the **Q** statement, then, from the definition of a conditional statement, **Table 1.4**, it follows that all the theorems resulting from true statements must themselves be true. Thus, whenever a model can be established for an axiomatic system, its axioms and the theorems which they imply will all be true, and the system will be **relatively consistent**; that is, there is a very high degree of likelihood for consistency.

To conclude this study of the property of consistency, let us examine some interpretations for the axiomatic system which you studied in Chapter 2. In order to facilitate your work, the axioms from Chapter 2 are listed below:

Axiom 1a: If A and B are any two points, then there exists at least one line containing both A and B.

Axiom 1b: If A and B are any two points, then there exists at most one line containing both A and B.

Axiom 2: If m is a line, then there exist at least three points on m.

Axiom 3: There exists at least one line.

Axiom 4: If m is a line, then there exists at least one point which is not on m.

Axiom 5a: If m is a line and A is a point not on it, then there exists at least one line n which contains A but has no point in common with m.

Axiom 5b: If m is a line and A is a point not on it, then there exists at most one line n which contains A but has no point in common with m.

Axiom 6: If m is a line, then there exist at most three points on m.

Since these axioms were based on a set which was presented initially by J. W. Young, the system composed of Axioms 1–6 and Theorems 1–14 will be denoted as System **Y**. As you recall, only five axioms were used to establish Theorems 1–9. Hence Axioms 1–5 and Theorems 1–9 can be considered as a **subsystem** which will be denoted as System **Y***.

Example 1: Is the following interpretation, in which "point" means "number" and "line" means "column of numbers," a model for System **Y***?

$$1$$
$$2$$
$$3$$

Solution: No, since Axiom 4 is false for this interpretation. ∎

Example 2: Is the following interpretation, in which "point" means "letter" and "line" means "column of letters," a model for System **Y***?

$$A \quad D$$
$$B \quad E$$
$$C$$

Solution: No, for at least one of Axioms 1–5 is false. In particular, Axiom 2 is false since the second column contains only two letters. ∎

A natural question which should have occurred to you by now is, *How many axioms need to be shown false when an interpretation is being examined?* Since a model results only when all axioms become true for a particular interpretation, it is sufficient to show that any *one* axiom is false.

Example 3: Is the following interpretation, in which "point" means "person" and "line" means "committee," a model for System **Y**?

Finance: George   Dave   Bill
Social: Dave   Andy   Ivan
Membership: Craig   Bill   Andy
Education: Frank   Andy   Hank
Program: Dave   Craig   Hank
Judicial: Frank   George   Craig
Bylaws: Ivan   George   Hank
Athletics: Bill   Frank   Ivan
Scholarship: Craig   Ivan   Ed
Elections: Dave   Ed   Frank
Revenue: Ed   George   Andy
Service: Bill   Ed   Hank

Solution: Yes, since it causes each of Axioms 1–6 to be true. ∎

Exercises 3.2

1. Is the following interpretation, in which "point" means "number" and "line" means "column of numbers," a model for System **Y**?

4	1	2	2	9	2	5	2	8	7	8	8
3	6	1	5	5	4	1	7	3	5	7	1
9	7	3	8	6	6	4	9	6	3	4	9

2. Is the following interpretation a model for System **Y***?

   a single point; no line.

3. If an interpretation is a model for a system, is it also a model for a subsystem of that system? Why?

The following set of axioms is for use in problems 4–8. Axioms 1–5 can be used to prove theorems in a system which will be called System **Z***, while Axioms 1–6 can be used to prove theorems in a system which will be called System **Z**.

Axiom 1a: If A and B are two points, then there exists at least one line containing both A and B.

Axiom 1b: If A and B are two points, then there exists at most one line containing both A and B.

Axiom 2: If m is a line, then there exist at least three points on m.

Axiom 3: There exists at least one line.

Axiom 4: If m is a line, then there exists at least one point not on m.

Axiom 5: If m and n are two lines, then there exists at least one point belonging to both m and n.

Axiom 6: If m is a line, then there exist at most three points on m.

4. Is the following interpretation, in which "point" means "number" and "line" means "column of numbers," a model for System **Z**?

$$\begin{array}{cccc} 1 & 4 & 1 & 1 \\ 2 & 5 & 4 & 5 \\ 3 & 6 & 7 & 8 \end{array}$$

5. Is the following interpretation, in which "point" means "letter" and "line" means "column of letters," a model for System **Z**?

$$\begin{array}{cccccccc} D & A & B & C & A & A & B \\ E & B & F & D & E & C & C \\ G & D & G & F & F & G & E \end{array}$$

6. Is the following interpretation, in which "point" means "number" and "line" means "column of numbers," a model for System **Z**?

$$\begin{array}{ccccccc} 4 & 3 & 2 & 1 & 2 & 1 & 1 \\ 6 & 5 & 4 & 3 & 3 & 2 & 5 \\ 7 & 6 & 5 & 4 & 7 & 6 & 7 \end{array}$$

7. Is the following interpretation, in which "point" means "person" and "line" means "team," a model for System **Z***?

   Blue: Jim Dan Melinda
   Orange: Dan Betty Carol
   Gold: Karen Betty Nan Rebecca
   Garnet: Jim Nan Craig Melinda Bob

8. Is the following interpretation, in which "point" means "letter" and "line" means "column of letters," a model for System **Z***?

$$\begin{array}{ccccccccccc} H & K & B & A & C & C & G & B & L & J & A & F & H \\ M & M & I & B & I & J & M & C & A & L & E & L & K \\ L & F & G & D & A & G & E & E & G & I & F & D & I \\ B & J & F & J & M & H & D & K & K & E & H & C & D \end{array}$$

9. Develop a model for a System **S**, whose axioms are the following:
   Axiom 1: There exist exactly three points.

54 CHAPTER 3

Axiom 2: If A and B are two points, they lie together on at least one line.

Axiom 3: If A and B are two points, they lie together on at most one line.

Axiom 4: Not all points are on the same line.

Axiom 5: If m and n are two lines, they have exactly one point in common.

10. Develop a model for a System **H** which can be described as:

    Let S be a set of people, with the members of S designated as A, B, C, ... Certain subsets of S are called **committees**, an undefined term.

    Definition: Two committees are **distinct** if one of them has a member which is not a member of the other.

    Definition: Two committees are **unrelated** if they have no member in common.

    Axiom 1: S contains at least two people.

    Axiom 2: No committee is without a member.

    Axiom 3: Each pair of people in S serve together on exactly one committee.

    Axiom 4: For each committee, there exists exactly one unrelated committee.

11. Is the following interpretation, in which "person" means "letter" and "committee" means "set," a model for System **H** of problem 10?

    $\{A,E\}$ $\{B,C,E\}$ $\{A,B,C,D\}$ $\{B,D,E\}$

12. A manufacturer of costume jewelry has designed a piece which is composed of colored beads placed on wires as indicated in the following diagram. In this jewelry, there are seven beads: amber, burgundy, coral, fuchsia, khaki, mauve, and turquoise. These beads are arranged on seven wires in such a way that each bead is on three wires and each wire contains three beads.

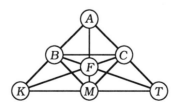

PROPERTIES OF AN AXIOMATIC SYSTEM   55

Is this interpretation, in which "point" means "bead" and "line" means "wire," a model for System Z^*?

$w_1 = \{A,B,K\}$ $w_2 = \{A,C,T\}$ $w_3 = \{B,F,T\}$ $w_4 = \{C,F,K\}$
$w_5 = \{K,M,T\}$ $w_6 = \{A,F,M\}$ $w_7 = \{B,C,M\}$

13. Is the bead-wire interpretation in problem 12 a model for System H of problem 10?

14. Consider the following interpretation, in which "point" means "letter" and "line" means "column of letters."

V	F	H	J
F	C	F	C
J			

Is this interpretation a model for the system which has the following axiom set? If your answer to this question is *no*, how can you alter the interpretation, by adding columns, so that it *will* be a model? Provide a reason for each new column which you add. When you add a column, be certain it does not cause another axiom to become false.

$S = \{C,F,H,J,V\}$ **Undefined terms**: point, line

Definition: Two lines are said to be parallel if they have no point in common.
Axioms 1: If m is a line, it contains at least two points.
Axiom 2: For each line, there exists at least one parallel line.
Axiom 3: If m is a line, there exists a point not on m.
Axiom 4: If A and B are any two points, they lie together on exactly one line.
Axiom 5: There exist at least two members in S.

3.3 Independence of Axioms

Often when a set of axioms is being established for an axiomatic system, the primary concern relates to the consistency of the resulting system. As you learned in the preceding section, this, in fact, *must* be the primary concern since consistency is absolutely essential.

After axioms have been selected, however, it is frequently considered desirable to refine the collection of axioms in order that only **primitive** statements are included. That is to say, only those axioms which cannot be derived from one or more of the other axioms are retained in the set, while the others are restated as theorems.

Definition 5: An axiom is called **independent** in a set of axioms if it cannot be proved with the use of one or more of the other axioms of the set.

Test for Independence: If **S** is a consistent axiomatic system and A is an axiom in **S**, then A is **independent** in the set of axioms for **S** if and only if the system which results when A is replaced with not-A is also consistent.

The rationale for this test is reasonably fundamental. If **S** is consistent and A is not independent, and hence a theorem, then A is implied by the other axioms. Thus, not-A and A would both be theorems in **S**, and, as a result, **S** would, in fact, *not* be consistent.

Frequently, when an axiom of a consistent system is being checked for independence, it is not actually replaced with its negation before a model is developed. Instead, an interpretation is determined which causes the axiom being tested to be false while all the other axioms are true. Since a statement and its negation always have opposite truth values, this approach produces the same result as that described in the **Test for Independence**. The primary reasons for using this equivalent alternative technique are that it is less time-consuming and that the negation of a statement is sometimes a cumbersome statement to use.

In problems 1 and 3 of **Exercises 3.2**, it was established that System **Y*** is consistent, so axioms for this system can be examined for examples.

Example 4: Is Axiom 1a independent in System **Y***?

Solution: The interpretation

$$m = \{A,B,C\} \qquad n = \{D,E,F\}$$

causes Axiom 1a to be false, while all the other axioms in **Y*** remain true. Hence, Axiom 1a is independent. ∎

Example 5: Is Axiom 1b independent in System **Y***?

Solution: The interpretation

1	1	1	1	1	2	2	2	3	4
2	2	3	3	4	3	4	4	4	5
6	3	5	6	5	6	5	6	5	6

causes Axiom 1b to be false, while all the other axioms in **Y*** remain true. Hence, Axiom 1b is independent. ■

Example 6: Is Axiom 3 independent in System **Y***?

Solution: The interpretation

a single point; no line

causes Axiom 3 to be false, while all the other axioms in **Y*** remain true. (You may wish to examine **Table 1.4** to be reminded of the truth value of an **if P, then Q** statement when **P** is false.) Hence, Axiom 3 is independent. ■

A concluding observation is in order concerning the necessity and desirability of axioms being independent. While it is often considered esthetically desirable for all the axioms of a system to be independent, it is sometimes impractical. It is not at all uncommon for a theorem whose proof is exceptionally difficult to be assumed and used as an axiom in order that subsequent related theorems can be proved.

Exercises 3.3

1. Can the following interpretation, in which "point" means "letter" and "line" means "column of letters," be used to show the independence of Axiom 6 of System **Y** of **3.2**?

A	A	C
D	C	B
F	E	D
	B	F

2. Can the following interpretation, in which "point" means "number" and "line" means "set," be used to show the independence of Axiom 2 of System **Z** of problem 3 of **Exercises 3.2**?

    {2,3,5}   {1,4,5}   {2,4}   {1,3}   {1,2}

3. Develop interpretations to show that each of Axioms 2, 4, 5 is independent in System **Y*** of **3.2**.

4. Consider the following interpretation, in which "point" means "letter" and "line" means "column of letters."

R	Q	P	T	R
P	T	R	P	Q
T	S		S	
			R	

Can this interpretation be employed to show the independence of *any* axiom in the system for which the following hold? Explain your answer. *You may use the fact that the following system is consistent.*

Undefined terms: point, line

Axioms 1: There exist exactly five points.

Axiom 2: If A and B are any two points, they are together on at least one line.

Axiom 3: Not all points are on the same line.

Axiom 4: If m and n are any two lines, they have at least one point in common.

5. In problem 9 of **Exercises 3.2**, you showed that System **S** is consistent.
   (a) Develop interpretations to show that all but one of the axioms in **S** are independent.
   (b) Use the independent axioms of **S** to prove the axiom which is not independent.

6. In problem 10 of **Exercises 3.2**, you showed that System **H** is consistent. Are the axioms of **H** independent?

3.4 Completeness of a System

The least important of the three characteristics associated with an axiomatic system is that of **completeness**. As the name implies, when a system possesses this characteristic, there is often little additional information of interest which can be added to the system.

Definition 6: An axiomatic system **S** is **complete** if it is impossible to add an independent axiom to **S**.

Just as in the case of consistency and independence, there is a test for completeness. Before this test can be considered, however, some additional definitions are needed.

Definition 7: A pairing of the elements of two sets S_1 and S_2 is a **one-**

to-one correspondence if each element of S_1 is matched with exactly one element of S_2 and each element of S_2 is matched with exactly one element of S_1.

Definition 8: A one-to-one correspondence between two sets is said to **preserve relations** if every true statement made about elements in one set is also true about the corresponding elements in the other set.

Definition 9: Two models of an axiomatic system **S** are said to be **isomorphic** with respect to **S** if there exists at least one relations-preserving one-to-one correspondence between the elements of the two models.

Definition 10: An axiomatic system **S** is called **categorical** if each pair of models for **S** is isomorphic with respect to **S**.

Test for Completeness: If an axiomatic system **S** is categorical, then it is complete.

As was true in the case of the test for independence of an axiom within a system, this test for completeness is reasonably easy to understand. In fact, it can be explained in the form of an indirect argument.

If a system **S** is assumed to be categorical but not complete, then it would be possible for an independent axiom A to be added to the initial set S of axioms for **S**. Since A is independent, then models M_1 and M_2 could be developed for **S**, based on the axiom sets $S \cup \{A\}$ and $S \cup \{not - A\}$, respectively. Since **S** is categorical, then there must exist at least one relations-preserving one-to-one correspondence between the elements of M_1 and M_2. This would be impossible, however, since A is true regarding M_1 but false relative to M_2. Thus, the initial assumption is incorrect, and the statement in the test is true.

For examples and exercises relative to the characteristic of completeness, references will be made to systems and models which were considered in **3.2** and **3.3** and the related exercise sets. As Definition 10 indicates, it is not actually possible *to prove* that a system is categorical, for it would be virtually impossible to know that all possible models for a system had been developed. Hence, in the examples and exercises only the idea of an isomorphism will be examined.

Example 7: Determine an isomorphism between the model from Example 3 and the model from problem 1 of **Exercises 3.2**.

Solution: The two models to be considered are:

m	**Finance**:	George	Dave	Bill							
n	**Social**:	Dave	Andy	Ivan							
o	**Membership**:	Craig	Bill	Andy							
p	**Education**:	Frank	Andy	Hank							
q	**Program**:	Dave	Craig	Hank							
r	**Judicial**:	Frank	George	Craig							
s	**Bylaws**:	Ivan	George	Hank							
t	**Athletics**:	Bill	Frank	Ivan							
u	**Scholarship**:	Craig	Ivan	Ed							
v	**Elections**:	Dave	Ed	Frank							
w	**Revenue**:	Ed	George	Andy							
x	**Service**:	Bill	Ed	Hank							

m^*	n^*	o^*	p^*	q^*	r^*	s^*	t^*	u^*	v^*	w^*	x^*
4	1	2	2	9	2	5	2	8	7	8	8
3	6	1	5	5	4	1	7	3	5	7	1
9	7	3	8	6	6	4	9	6	3	4	9

As C^* indicates, below, the one-to-one correspondence C preserves the relationships existing between the corresponding elements of the two models.

C

					C^*			
Andy	–	1	Frank	–	8			
Bill	–	2	George	–	6			
Craig	–	3	Hank	–	9			
Dave	–	4	Ivan	–	5			
Ed	–	7						

m	–	r^*	s	–	q^*
n	–	s^*	t	–	p^*
o	–	o^*	u	–	v^*
p	–	x^*	v	–	w^*
q	–	m^*	w	–	n^*
r	–	u^*	x	–	t^*

Hence, the two models are isomorphic. ∎

Example 8: Why is the following correspondence between elements of the two models discussed in Example 7 *not* an isomorphism?

Andy	–	1	Bill	–	2
Craig	–	3	Dave	–	4
Ed	–	7	Frank	–	8
George	–	9	Hank	–	5
Ivan	–	6			

Solution: Committee n contains Andy, Dave, Ivan, which correspond to 1, 4, 6, respectively, but there is no column containing these numbers. Hence, this is not an isomorphism. ∎

Exercises 3.4

1. Determine an isomorphism for the following two arrays. Give both the one-to-one correspondence of the elements and the correspondence of the columns.

   I. C F E B B   II. 2 8 1 5 3
       E A A D F   5 4 3 8 2
       D C D F A   3 1 8 4 4

2. Set up an isomorphism between the following model for System **Y** of **3.2** and the model in problem 1 of **Exercises 3.2**.

   H H H A E C G A F I H A
   A E C E G B I C D E B D
   I D G B F D B F I C F G

3. Determine an isomorphism between the models in problems 5 and 6 of **Exercises 3.2**.

4. Consider the following interpretation for System **Y** of **3.2**:

   1 1 1 1 2 2 2 3 3 3 4 5
   2 4 6 8 4 5 6 4 5 7 7 6
   3 5 7 9 9 7 8 6 8 9 8 9

   (a) Establish that this interpretation is a model for System **Y**.
   (b) Determine an isomorphism between this model and the model in Example 3.

5. Consider the following collection of statements:
   1. If A and B are two points, they determine exactly one line.
   2. For any given line, there exists at least one point which is not on the line.
   3. If m and n are two lines, they have exactly one point in common.

   If you were to add exactly *one* more statement to this list in order to obtain a set of axioms for an axiomatic system, what would it be? Why? *Think before you write.* There will likely be two statements which come to your mind. Add the one which will allow you the *most* use of these three statements.

   ****

SUPPLEMENTARY TOPICS

CONSTRUCTIONS

10. Using Construction 9 from **Appendix 1** and an initial segment measuring 1 inch, construct a segment whose length will be $\sqrt{13}$ inches.

11. Using Construction 22 from **Appendix 1** and an initial segment measuring 1 inch, construct a segment whose length will be $\sqrt{6}$ inches.

12. Draw two intersecting lines and construct a circle which is tangent to each of these lines and has a radius whose measure is the same as that of the given segment $\overline{AB}$. How many such circles are possible?

$$\overline{}$$
$$\ \ A \qquad\qquad B$$

13. Construct a triangle $\triangle ABC$ such that $\angle A$ is congruent to the given angle $\angle D$, side $\overline{AB}$ is congruent to the given segment $\overline{EF}$, and the altitude from vertex C is congruent to the given segment $\overline{GH}$.

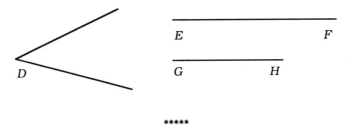

*****

PROBLEMS

6. A little girl's tricycle has a front wheel whose diameter is 20 inches and rear wheels whose diameters are 8 inches. As she rides down the sidewalk one afternoon, she is pedaling the front wheel at 30 revolutions per minute. How many revolutions per *second* are the rear wheels making?

7. Determine the area and perimeter of the shaded region in the diagram below. Points O, O^*, and A are collinear, with O and O^* being the centers of circles, each having radius 2 units.

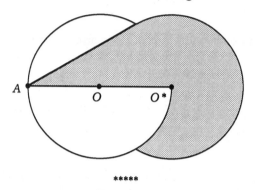

*****

TRANSFORMATION GEOMETRY

T-3: Line Reflections

Definition T-7: If m is a line in a plane, then a **reflection R in m** is a transformation of the plane such that

$$\mathbf{R}_m(P) = \begin{cases} P' = P, \text{ if } P \in m \\ P', \text{ where } m \text{ is the perpendicular} \\ \text{bisector of the segment } \overline{PP'}, \text{ if } P \notin m. \end{cases}$$

Remark T-1: A **reflection in a line** is often called a **line reflection**.

Theorem T-5: A line reflection is an isometry.

Example T-8: When a polygon is reflected in a line, its image can be found by reflecting its vertices and sketching the polygon which they determine. (See **Figure T-10**.) ∎

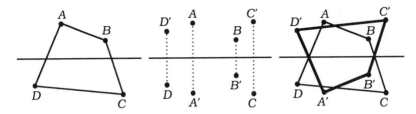

Figure T-10

Example T-9: If m is a line, $P = (-4,2)$, and $\mathbf{R}_m(P) = P' = (2,6)$, what is an equation for line m?

Since m must serve as the bisector of the segment $\overline{PP'}$, m contains the midpoint of the segment: $\left(\frac{-4+2}{2}, \frac{2+6}{2}\right) = (-1, 4)$.

Since m must be perpendicular to $\overline{PP'}$, then m has slope

$$-\frac{1}{(6-2)/(2+4)} = -\tfrac{3}{2}.$$

Hence, an equation for m is $y - 4 = -\tfrac{3}{2}(x+1)$, which simplifies to $y = -\tfrac{3}{2}x + \tfrac{5}{2}$. ∎

Example T-10: If C_1 and C_2 are circles and m is a line, as shown in **Figure T-11**, determine a segment $\overline{PQ}$ such that $P \in C_1$, $Q \in C_2$, and m is the perpendicular bisector of $\overline{PQ}$.

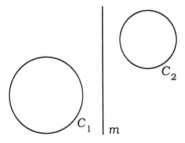

Figure T-11

Find $\mathbf{R}_m(C_2)$, the image of C_2, by reflecting the center of C_2 across m and using the radius of C_2 to draw C_2'. (See **Figure T-12**.)

Label a point where C_1 and C_2' intersect as P. When P is reflected across line m, $\mathbf{R}_m(P)$ is the point of C_2 which is the preimage of P; label $\mathbf{R}_m(P)$ as Q. (See **Figure T-13**.)

SUPPLEMENTARY TOPICS

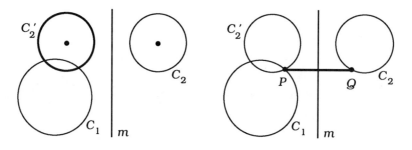

Figure T-12   **Figure T-13**

Line m is the perpendicular bisector of $\overline{PQ}$. (Note that a second such segment can be drawn, if the other point of $C_1 \cap C_2$ is selected to be P.) ■

Example T-11: If P and Q are points and m is a line, as shown in **Figure T-14**, determine the shortest path from P to a point on m to Q.

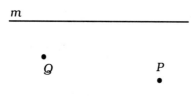

Figure T-14

Find $\mathbf{R}_m(Q) = Q'$, and draw the line determined by Q' and P. (See **Figure T-15**.) Label as T the point of intersection of the two lines.

Draw segment $\overline{QQ'}$, and label its point of intersection with m as S. Draw segment $\overline{QT}$. (See **Figure T-16**.)

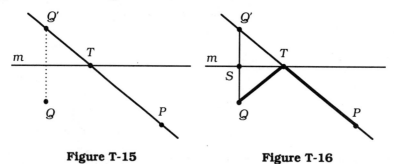

Figure T-15   **Figure T-16**

Since $\mathbf{R}_m(Q) = Q'$, then $\overline{SQ'} \cong \overline{SQ}$ and each of $\angle TSQ'$ and $\angle TSQ$ is a right angle. Also, $\overline{ST} \cong \overline{ST}$. Thus, $\triangle TSQ' \cong \triangle TSQ$, by Side Angle Side.

The shortest path from P to Q' is segment $\overline{PQ'}$. Since $m\overline{PQ'} = m\overline{PT} + m\overline{TQ'} = m\overline{PT} + m\overline{TQ}$, the shortest path from P to a point on m to Q will be from P to T to Q. ∎

A physics principle related to the reflection of a light ray by a mirrored surface can be adapted for use with line reflections:

Light Principle: If a ray of light is reflected by a flat, mirrored surface, the angle formed by the surface and the reflection will be congruent to the angle formed by the surface and the light ray. These two angles are known, respectively, as the **angle of reflection** and the **angle of incidence**. (See **Figure T-17**.)

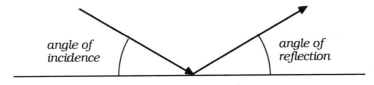

Figure T-17

When a ball having no spin is bounced off a flat wall, its motion is similar to that of a light ray in that its "angle of reflection" is congruent to its "angle of incidence." A reexamination of **Figure T-16** will be helpful in understanding how this relates to line reflections. Since $\triangle TSQ' \cong \triangle TSQ$, then $\angle 1 \cong \angle 2$. Since $\angle 1$ and $\angle 3$ are vertical angles, then $\angle 1 \cong \angle 3$. Thus, $\angle 2 \cong \angle 3$.

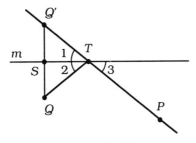

Figure T-16A

Bounce Principle: A ball, having no spin, will travel from a point A to a point B by bouncing off a straight wall, if the ball is rolled from A toward the point at which B would be reflected by the wall.

Example T-12: A 4-foot by 8-foot pool table with a cue ball located at $C(2,1)$ and a second ball located at $A(6,3)$ is shown in **Figure T-18**. At what

SUPPLEMENTARY TOPICS

point on the $\overline{(0,0)(0,4)}$ side of the table should the cue ball be aimed, if it has no spin and is to bounce off each of the $\overline{(0,0)(0,4)}$ and $\overline{(0,4)(8,4)}$ walls before striking ball A? How far will the cue ball travel before striking ball A?

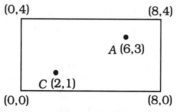

Figure T-18

The reflection of $A(6,3)$ across the line containing $(0,4)$ and $(8,4)$ produces $A'(6,5)$. When $A'(6,5)$ is, in turn, reflected across the line containing $(0,0)$ and $(0,4)$, the result is $A''(-6,5)$. (See **Figure T-19**.) Hence, the cue ball should be aimed at the point $A''(-6,5)$.

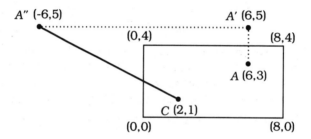

Figure T-19

Since the point $A''(-6,5)$ obviously is not located on the pool table, it will be necessary to determine the aiming point associated with a table point.

The line CA'', containing $C(2,1)$ and $A''(-6,5)$, has slope $(5-1)/(-6-2) = -\frac{1}{2}$. Hence, an equation for it is $y - 1 = -\frac{1}{2}(x - 2)$, which simplifies to $y = -\frac{1}{2}x + 2$. The point at which the line CA'' intersects $\overline{(0,0)(04)}$ is $T(0,2)$, obtained by substituting 0 for x in $y = -\frac{1}{2}x + 2$. (See **Figure T-20**.)

Thus, the cue ball should be aimed at the point $T(0,2)$ on the $\overline{(0,0)(04)}$ side.

The line TA', containing $T(0,2)$ and $A'(6,5)$, has slope $(5-2)/(6-0) = \frac{1}{2}$. Hence, an equation for it is $y - 2 = \frac{1}{2}(x - 0)$, which simplifies to $y = \frac{1}{2}x + 2$.

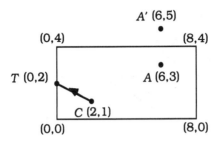

Figure T-20

The point at which the line TA' intersects $\overline{(0,4)(8,4)}$ is $S(4,4)$, obtained by substituting 4 for y in $y = \tfrac{1}{2}x+2$. (See **Figure T-21**.)

Thus, the cue ball will bounce off the $\overline{(0,4)(8,4)}$ wall at the point $S(4,4)$. (See **Figure T-22**.)

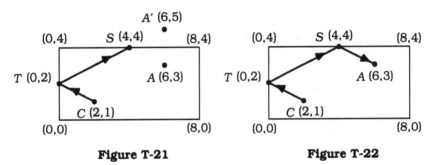

Figure T-21   **Figure T-22**

Since the ball will travel along three line segments, the total distance which it will travel will equal the sum of the measures of these segments:

$$\sqrt{(0-2)^2+(2-1)^2} + \sqrt{(4-0)^2+(4-2)^2} + \sqrt{(6-4)^2+(3-4)^2}$$

$$= \sqrt{4+1} + \sqrt{16+4} + \sqrt{4+1} = 4\sqrt{5}.$$

Thus, the cue ball will travel a total of $4\sqrt{5}$ units before striking ball A. ∎

Exercises T-3

1. Using a straightedge, draw a diagram similar to the one in **Figure T-23** and construct the image of $\triangle ABC$ when it is reflected in line m.

SUPPLEMENTARY TOPICS   69

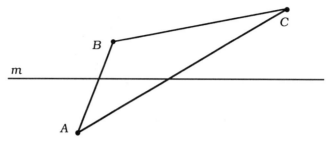

Figure T-23

2. In **Figure T-23**, above, the points A, B, and C are arranged in a clockwise order. Hence, it can be stated that the points A, B, C have a clockwise **orientation**. Examine $\triangle A'B'C'$, which you constructed in problem 1, and determine the orientation of points A', B', C'. Does a line reflection preserve or reverse orientation?

3. In **Figure T-24**, A and B represent two cities which are located along a straight river m. Sketch a similar diagram on an appropriately scaled coordinate system, and use algebraic calculations to determine a point W on m at which a jointly-owned water-pumping station should be located, if the minimum amount of pipe is to be used.

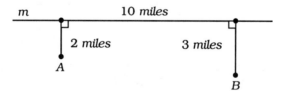

Figure T-24

4. The diagram in **Figure T-25** represents a 4-foot by 9-foot pool table having only corner pockets. If a ball having no spin leaves the $(4,4)$ point, strikes the opposite side at the $(6,0)$ point, and continues traveling until it enters a pocket, which pocket will it enter?

70   CHAPTER 3

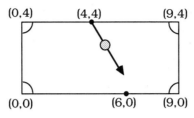

Figure T-25

5. Determine the missing entry in each row of the following table:

	equation of line m	point P	$\mathbf{R}_m(P) = P'$
(a)	$y = x$	$(2,-5)$	?
(b)	$y = 0$	?	$(-3,-5)$
(c)	?	$(-3,7)$	$(3,7)$
(d)	$x = 2$	$(4,0)$	?
(e)	?	$(1,3)$	$(1,-5)$
(f)	?	$(4,-2)$	$(2,-4)$
(g)	$y = 3x - 2$	$(0,5)$	?
(h)	?	$(2,3)$	$(26/5, 7/5)$

6. Construct a diagram similar to that shown in **Figure T-26**, and construct a square $DEFG$, having $D \in C_1$, $F \in C_2$, $E \in m$, and $G \in m$. How many such squares are possible?

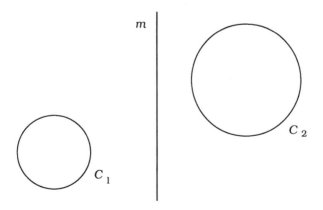

Figure T-26

7. Using an appropriately scaled coordinate system, develop three diagrams similar to that pictured in **Figure T-27**. Assuming the diagram represents a hole on a miniature golf course, use constructions to show a path which a non-spinning ball must follow, if a hole-in-one is to be made in each of the following cases. In addition, use algebraic calculations to determine the name of the first wall point at which the ball must bounce.
   (a) The ball bounces off only one wall.
   (b) The ball bounces off two walls.
   (c) The ball bounces off three walls.

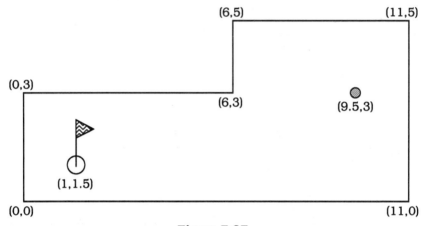

Figure T-27

8. In the diagram in problem 7, would it be possible for the ball to bounce off four walls and still go in the hole? If not, why?

9. In **Figure T-28**, lines *m* and *n* represent parallel walls. Construct a diagram similar to this one, and construct a path which a non-spinning ball could follow, if it is to leave point *P* and travel to point *Q* along a path which involves two bounces off wall *m* and one bounce off wall *n*.

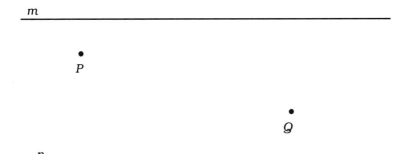

Figure T-28

10. Develop an informal proof for Theorem T-5.

PART II

An Axiomatic Development of Elementary Geometry

Geometry as we know it today is basically little different from Euclid's *Elements*. However, numerous mathematicians since have done research in geometry, often in attempts to establish that Euclid's axiom relating to parallel lines is not an independent axiom or to eliminate some of the logical defects of the *Elements*.

The end results of the research of several mathematicians exploring Euclid's axioms were systems differing significantly from that described by Euclid. Ultimately, these geometric systems were characterized as being non-Euclidean. Some of the mathematicians making significant observations and contributions were the Greek Proclus (410-485), the Italian G. Saccheri (1667-1733), the Swiss J. Lambert (1728-1777), the German F. Gauss (1777-1855), the Hungarian J. Bolyai (1802-1860), the Russian N. Lobachevsky (1793-1856), the Germans B. Riemann (1826-1866), F. Klein (1849-1925), and M. Pasch (1843-1930), the Italian G. Peano (1858-1932), and the German David Hilbert (1862-1943).

Hilbert's works in geometry, sometimes described as being non-Euclidean in nature, are, in fact, *refinements* of Euclidean geometry. In his 1899 book *Foundations of Geometry*, Hilbert presented a set of axioms which eliminate some of the logical defects of Euclidean geometry plaguing mathematicians since the appearance of the *Elements*. Although it was not the first work presenting geometry in an axiomatic manner, *Foundations* is generally regarded as the first to display the axiomatic method in its modern form. In addition to presenting axioms and theorems in a strictly logical development, Hilbert presented materials concerning the axiomatic system properties of consistency, independence, and completeness.

Hilbert's work provides the basis for the development of elementary geometry in this part of this book. As in Chapter 2, definitions, axioms, and theorems, as well as language rules and notation rules, will be used to introduce undefined terms and symbolisms. In an effort to provide for a reasonable pace and a realistic amount of rigor, some non-independent axioms will be used. Thus, the emphasis will be placed on the development of a logical system which will lay the foundations of a geometry similar in many respects to that of Euclid.

Unlike Chapter 2, in which proofs were presented for all of the theorems, many of the theorems in this part of the book will be presented without proof in order that you will have opportunities to develop proofs.

Chapter 4: The Foundations of Plane Geometry

4.1 Introduction

The material in this chapter will relate to characteristics of subsets of a plane, with no reference intended, either explicitly or implicitly, to subsets of space. A later chapter, "The Foundations of Space Geometry," will be devoted to a short study of subsets of space. Since no reference will be made to space, it can be assumed, in this and subsequent chapters pertaining to plane geometry, that there is only one plane. This assumption will create no major difficulties in the later chapter concerning subsets of space, for in that chapter every plane will possess the same characteristics as those which will be established for this one plane.

4.2 Existence and Incidence

The axioms and theorems of this section establish connections between the existence and incidence of points and lines on a plane.

Language Rule 1: Point, line, and **plane**, and the vocabulary of logic and sets will be undefined terms.

Language Rule 2: Points will be considered the elements of **linear geometry**; points and lines will be considered the elements of **plane geometry**.

Notation Rule 1: Points will be denoted with upper case letters, lines will be denoted with lower case letters, and planes will be denoted with boldface upper case letters.

Language Rule 3: When numbers are used in the statement of axioms and theorems, they will describe distinct objects.

Language Rule 4: Statements such as *m* **contains** *A*, *A* **lies on** *m*, and *m* **passes through** *A* will be used to indicate that a point *A* is an element of a line *m*.

Language Rule 5: Statements such as *m* **and** *n* **have** *A* **in common** and *m* **and** *n* **intersect at** *A* will be used to indicate that a point *A* is an element of two lines. Similar statements will be used when a point *A* is an element of more than two lines.

Notation Rule 2: The symbol "=" will be used to represent "identical."

Since it is to be assumed that there is only one plane, which contains all points, then this assumption must be formally stated as an axiom. In order that it can be easily discarded in the later chapter dealing with subsets of space, this axiom will be numbered zero.

Axiom 0: There exists exactly one plane, which contains all points.

Axiom 1a: If A and B are two points, then there exists at least one line containing them.

Axiom 1b: If A and B are two points, then there exists at most one line containing them.

Language Rule 6: A line m which contains two points A and B can be described as having been **determined by A and B**.

Notation Rule 3: If A and B are any two points on a line m, then m may also be denoted as AB or BA.

Axiom 2: If m is a line, then it contains at least two points.

Axiom 3: If m is a line, then there exists at least one point which does not lie on it.

Axiom 4: There exist at least two points.

Theorem 1: There exist at least three points.

Theorem 2: Two lines have at most one point in common.

Definition 1: Two or more points are called **collinear** if they lie on the same line.

Theorem 3: If A and B are two points on a line m and C is a point not on m, then A, B, C are noncollinear and there exist lines n and k, containing A, C and B, C, respectively, which are distinct from m and from each other.

Theorem 4: There exist at least two lines through any point.

Theorem 5: If A is a point, then there exists at least one line which does not contain it.

Exercises 4.2

1. Develop a proof for Theorem 1.

2. Develop a proof for Theorem 2.

3. Develop a proof for Theorem 3.

4. Develop a proof for Theorem 4.

5. Develop a proof for Theorem 5.

6. Develop a finite model for the system described by the axioms of this section.

7. Attempt to develop a proof for each of the following statements, using any of the materials in this section other than Theorem 3. If you cannot do so, thoroughly explain the difficulty involved. (It might be helpful to employ diagrams to explain the difficulties.)
   (a) There exist at least four points.
   (b) There exist at least three lines.
   (c) There exist at least four lines.

4.3 Between—An Order Relation

The material in this section will establish some properties of an order relation which can be developed among the points on a line.

Language Rule 7: Between will be an undefined term.

Notation Rule 4: Point B is between points A and C will be denoted as $A \bullet B \bullet C$.

Axiom 5: If A and B are two points, then
(a) there exists at least one point C such that $A \bullet C \bullet B$;

(b) there exists at least one point D such that $A \bullet B \bullet D$;
(c) there exists at least one point E such that $E \bullet A \bullet B$. (See **Figure 4.1**.)

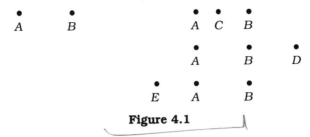

Figure 4.1

Axiom 6: If A, B, C are points such that $A \bullet B \bullet C$, then A, B, C are distinct and collinear.

Axiom 7: If A, B, C are points such that $A \bullet B \bullet C$, then $C \bullet B \bullet A$.

Axiom 8: If A, B, C are three collinear points, then exactly one of the following is true:
(a) $A \bullet B \bullet C$;
(b) $A \bullet C \bullet B$;
(c) $C \bullet A \bullet B$. (See **Figure 4.2**.)

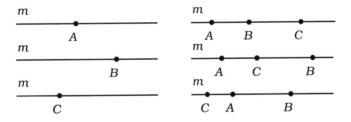

Figure 4.2

Axiom 9: If A, B, C, D are points such that $A \bullet B \bullet C \bullet D$, then $A \bullet B \bullet C$, $A \bullet B \bullet D$, $A \bullet C \bullet D$, and $B \bullet C \bullet D$. (Analogous statements can be made for more than four points.) (See **Figure 4.3**.)

Axiom 10: If A, B, C, D are four collinear points and $A \bullet B \bullet C$, then exactly one of the following is true:
(a) $A \bullet B \bullet C \bullet D$;
(b) $A \bullet B \bullet D \bullet C$;
(c) $A \bullet D \bullet B \bullet C$;
(d) $D \bullet A \bullet B \bullet C$. (See **Figure 4.4**.)

THE FOUNDATIONS OF PLANE GEOMETRY

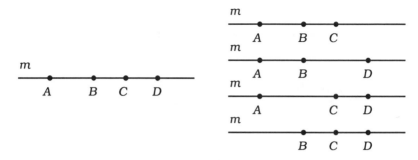

Figure 4.3

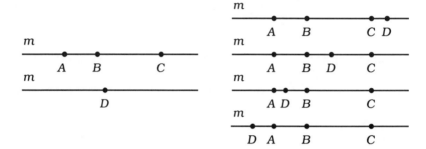

Figure 4.4

Theorem 6: If A, B, C, D are points such that $A \bullet B \bullet C$ and $B \bullet C \bullet D$, then $A \bullet B \bullet C \bullet D$.

Proof: Suppose A, B, C, D are points such that $A \bullet B \bullet C$ and $B \bullet C \bullet D$. By Axiom 6, A, B, C are distinct and collinear, and B, C, D are distinct and collinear. If A is D, then $A \bullet B \bullet C$ and $B \bullet C \bullet A$, which contradicts Axiom 8. Hence, A and D are distinct.

Since B and C are distinct, then, by Axiom 1a and Axiom 1b, there exists exactly one line m which contains them. Since A is collinear with B and C, then A is on m. Similarly, D is on m. Thus, A, B, C, D are collinear.

Thus, A, B, C, D are four collinear points, and $A \bullet B \bullet C$. By Axiom 10, exactly one of $A \bullet B \bullet C \bullet D$, $A \bullet B \bullet D \bullet C$, $A \bullet D \bullet B \bullet C$, and $D \bullet A \bullet B \bullet C$ is true.

If $A \bullet B \bullet D \bullet C$, $A \bullet D \bullet B \bullet C$, or $D \bullet A \bullet B \bullet C$ is true, then, by Axiom 9, $B \bullet D \bullet C$, $D \bullet B \bullet C$, or $D \bullet B \bullet C$, respectively. Each of these results would contradict Axiom 8, since $B \bullet C \bullet D$ was assumed true.

Hence, $A \bullet B \bullet D \bullet C$ is true. ∎

Theorem 7: If A, B, C, D are points such that $A \bullet B \bullet C$ and $A \bullet C \bullet D$, then $A \bullet B \bullet C \bullet D$.

Theorem 8: If A, B, C, D are points such that $A \bullet B \bullet D$ and $B \bullet C \bullet D$, then $A \bullet B \bullet C \bullet D$.

Theorem 9: If A, B, C, D are four points such that $A \bullet B \bullet C$ and $A \bullet B \bullet D$, then exactly one of $A \bullet C \bullet D$ and $A \bullet D \bullet C$ is true.

Proof: Suppose A, B, C, D are four points such that $A \bullet B \bullet C$ and $A \bullet B \bullet D$.

By Axiom 6, A, B, C are collinear and A, B, D are collinear. Since A and B are distinct, then, by Axiom 1a and Axiom 1b, there exists exactly one line m which contains them. Points C and D are also on m, so A, B, C, D are collinear.

Since A, B, C, D are four collinear points and $A \bullet B \bullet C$, then, by Axiom 10, exactly one of $A \bullet B \bullet C \bullet D$, $A \bullet B \bullet D \bullet C$, $A \bullet D \bullet B \bullet C$, and $D \bullet A \bullet B \bullet C$ is true.

If $A \bullet D \bullet B \bullet C$ or $D \bullet A \bullet B \bullet C$ is true, then, by Axiom 9, $A \bullet D \bullet B$ or $D \bullet A \bullet B$, respectively. Each of these two results would contradict Axiom 8, since $A \bullet B \bullet D$ was assumed.

Hence, exactly one of $A \bullet B \bullet C \bullet D$ and $A \bullet B \bullet D \bullet C$ is true. If $A \bullet B \bullet C \bullet D$ is true, then, by Axiom 9, $A \bullet C \bullet D$, and, if $A \bullet B \bullet D \bullet C$ is true, then, by Axiom 9, $A \bullet D \bullet C$ is true.

Thus, exactly one of $A \bullet C \bullet D$ and $A \bullet D \bullet C$ is true. ∎

Theorem 10: If A, B, C, D are four points such that $A \bullet B \bullet C$ and $A \bullet B \bullet D$, then exactly one of $B \bullet C \bullet D$ and $B \bullet D \bullet C$ is true.

Theorem 11: If A, B, C, D are four points such that $A \bullet B \bullet D$ and $A \bullet C \bullet D$, then exactly one of $A \bullet B \bullet C$ and $A \bullet C \bullet B$ is true.

Exercises 4.3

1. Develop a proof for Theorem 7.

2. Develop a proof for Theorem 8.

3. Develop a proof for Theorem 10.

4. Develop a proof for Theorem 11.

5. Develop a proof for this statement:
   If $A \bullet B \bullet C \bullet D$, then A, B, C, D are distinct and collinear.

6. Develop a proof for this statement:
   If m is a line, then it contains at least five points.

7. Develop a proof for this statement:
   If m is a line, then there exist at least four points which do not lie on it.

4.4 Halflines; Partition and Separation

The order relation described in the previous section can be used to prove some theorems relating to a type of subset of a line.

Definition 2: If m is a line and O and A are two points on m, then
$$L_1 = \{X \mid X \in m \text{ and } X \bullet A \bullet O\} \cup \{A\} \cup \{X \mid X \in m \text{ and } A \bullet X \bullet O\}$$
$$\text{and } L_2 = \{X \mid X \in m \text{ and } A \bullet O \bullet X\}$$
are called the **halflines** of m with respect to O relative to A. The point O is called the **endpoint** of each halfline. (See **Figure 4.5**.)

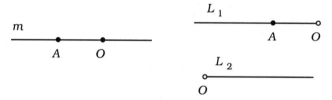

Figure 4.5

Language Rule 8: If O is a point on a line m, then the two halflines of m based on O can be described as having been determined by O and will be called the **sides of m relative to O** and **opposite halflines**. If B is a point on one of the halflines, then that halfline will be called the **B-side of O**. Two points lying on the same halfline will be described as being **on the same side of O**, while two points lying on different halflines will be described as being **on opposite sides of O**.

Notation Rule 5: If B is a point on a halfline having endpoint O, then the halfline may be denoted as $\overrightarrow{OB}$. Thus,

$$\overrightarrow{OB} = \{X \mid X \in OB \text{ and } O \bullet X \bullet B\} \cup \{B\} \cup \{X \mid X \in OB \text{ and } O \bullet B \bullet X\}.$$

Definition 3: If a set S can be expressed as the union of two or more nonempty, disjoint sets $S_1, S_2, \ldots$, then $S_1, S_2, \ldots$, are said to **partition** S, and the sets $S_1, S_2, \ldots$, are called a **partition** of S.

Theorem 12: If m is a line and O, A are two points on m, then $\{O\}$ and L_1, L_2, the halflines of m with respect to O relative to A, partition m.

Proof: Suppose m is a line and O, A are two points on m.
By Definition 2, the halflines of m with respect to O relative to A are
$$L_1 = \{X \mid X \in m \text{ and } X \bullet A \bullet O\} \cup \{A\} \cup \{X \mid X \in m \text{ and } A \bullet X \bullet O\}$$
$$\text{and } L_2 = \{X \mid X \in m \text{ and } A \bullet O \bullet X\}.$$

Claim: $\{O\}$ is nonempty.

Proof: $\{O\}$ contains point O, and, hence, is nonempty. ◆

Claim: L_1 is nonempty.

Proof: L_1 contains point A, and, hence, is nonempty. ◆

Claim: L_2 is nonempty.

Proof: By Axiom 5, there exists at least one point C such that $A \bullet O \bullet C$. By Axioms 6 and 1b, C is on m. Hence, L_2 is nonempty. ◆

Claim: $\{O\}$ and L_1 are disjoint.

Proof: Suppose $\{O\}$ and L_1 are not disjoint; that is, suppose $O \in L_1$. Since $O \in L_1$, then
$O \in \{X \mid X \in m \text{ and } X \bullet A \bullet O\}$, $O \in \{A\}$, or $O \in \{X \mid X \in m \text{ and } A \bullet X \bullet O\}$.
$O \notin \{A\}$, since O and A are two points.
$O \notin \{X \mid X \in m \text{ and } X \bullet A \bullet O\}$ and $O \notin \{X \mid X \in m \text{ and } A \bullet X \bullet O\}$, since $O \bullet A \bullet O$ and $A \bullet O \bullet O$ would contradict the distinctness part of Axiom 6.
Thus, $\{O\}$ and L_1 are disjoint. ◆

Claim: $\{O\}$ and L_2 are disjoint.

Proof: Analogous to the preceding **Claim**. ◆

Claim: L_1 and L_2 are disjoint.

Proof: Suppose Y is a point in L_2.
Since $Y \in L_2$, then $A \bullet O \bullet Y$.
Since $A \bullet O \bullet Y$, then, by Axiom 6, A and Y are distinct, and $Y \notin \{A\}$.
Since $A \bullet O \bullet Y$, then, by Axiom 8, $Y \bullet A \bullet O$ and $A \bullet Y \bullet O$ are not possible.
Hence, $Y \notin \{X \mid X \in m$ and $X \bullet A \bullet O\}$ and $Y \notin \{X \mid X \in m$ and $A \bullet X \bullet O\}$.
Thus, Y is not in L_1, and, hence, L_1 and L_2 are disjoint. ◆

Claim: $L_1 \cup \{O\} \cup L_2 \subseteq m$.

Proof: Suppose $Z \in L_1 \cup \{O\} \cup L_2$; that is, $Z \in L_1$, $Z \in \{O\}$, or $Z \in L_2$.
If $Z \in L_1$ or $Z \in L_2$, then $Z \in m$, by definition of L_1 and L_2. If $Z \in \{O\}$, then $Z = O$, which is on m.
Thus, $L_1 \cup \{O\} \cup L_2 \subseteq m$. ◆

Claim: $m \subseteq L_1 \cup \{O\} \cup L_2$.

Proof: Suppose $Z \in m$.
Since $Z \in m$, then $Z \in \{O, A\}$ or $Z \notin \{O, A\}$.
If $Z \in \{O, A\}$, then $Z = O$ or $Z = A$. Hence, $Z \in \{O\}$ or $Z \in L_1$, respectively.
If $Z \notin \{O, A\}$, then Z, O, A are distinct and collinear. By Axiom 8, exactly one of the following is true: $Z \bullet A \bullet O$, $A \bullet Z \bullet O$, $A \bullet O \bullet Z$.
If $Z \bullet A \bullet O$ or $A \bullet Z \bullet O$, then $Z \in L_1$, while, if $A \bullet O \bullet Z$, then $Z \in L_2$.
Hence, $m \subseteq L_1 \cup \{O\} \cup L_2$. ◆

Hence, L_1, $\{O\}$, L_2 are nonempty, disjoint sets whose union is m.
Thus, L_1, $\{O\}$, L_2 partition m. ∎

Definition 4: If S, S_1, S_2 are three nonempty, disjoint sets for which it is true that
(1) if X is an element of S_1 and Y is an element of S_2, then there exists an

element Z of S such that $X \bullet Z \bullet Y$; and
(2) if X, Y are two elements of S_1, then there does not exist an element Z of S such that $X \bullet Z \bullet Y$; and
(3) if X, Y are two elements of S_2, then there does not exist an element Z of S such that $X \bullet Z \bullet Y$;
then S is said to **separate** S_1 and S_2.

Theorem 13: If O is a point of a line m, then $\{O\}$ separates the two halflines which O determines on m.

Proof: Suppose O is a point on a line m. By Axiom 2, there exists a second point A on m.

By Definition 2, the halflines of m with respect to O relative to A are
$$L_1 = \{X \mid X \in m \text{ and } X \bullet A \bullet O\} \cup \{A\} \cup \{X \mid X \in m \text{ and } A \bullet X \bullet O\}$$
$$\text{and } L_2 = \{X \mid X \in m \text{ and } A \bullet O \bullet X\}.$$

Thus, $\{O\}$, L_1, L_2 partition m, by Theorem 12. Hence, $\{O\}$, L_1, L_2 are three nonempty, disjoint sets.

Claim: If X is an element of L_1 and Y is an element of L_2, then $X \bullet O \bullet Y$.

Proof: Suppose X is an element of L_1 and Y is an element of L_2.
Since $X \in L_1$, then $X \bullet A \bullet O$, $X = A$, or $A \bullet X \bullet O$. Since $Y \in L_2$, then $A \bullet O \bullet Y$.

If $X \bullet A \bullet O$ and $A \bullet O \bullet Y$ then, by Theorem 6, $X \bullet A \bullet O \bullet Y$. By Axiom 9, $X \bullet O \bullet Y$.

If $X = A$ and $A \bullet O \bullet Y$, then $X \bullet O \bullet Y$.

If $A \bullet X \bullet O$ and $A \bullet O \bullet Y$, then, by Theorem 7, $A \bullet X \bullet O \bullet Y$. By Axiom 9, $X \bullet O \bullet Y$.

Thus, if $X \in L_1$ and $Y \in L_2$, then $X \bullet O \bullet Y$. ◆

Claim: If X, Y are two elements of L_2, then $X \bullet O \bullet Y$ cannot occur.

Proof: Suppose X, Y are two elements of L_2.
Since $X \in L_2$ and $Y \in L_2$, then $A \bullet O \bullet X$ and $A \bullet O \bullet Y$. By Axiom 6, A, O, X are distinct and A, O, Y are distinct. Since X, Y are distinct, then A, O, X, Y are distinct points. Thus, by Theorem 10, exactly one of $O \bullet X \bullet Y$ and $O \bullet Y \bullet X$ is true. Hence, by Axiom 8, $X \bullet O \bullet Y$ cannot occur. ◆

Claim: If X, Y are two elements of L_1, then $X \bullet O \bullet Y$ cannot occur.

Proof: Analogous to the preceding **Claim**, with several parts. ◆

Thus, $\{O\}$ separates L_1 and L_2. ∎

Exercises 4.4

1. Develop a proof for this statement:

   If D is a point on line AB such that $D \in \overrightarrow{AB}$ and $D \in \overrightarrow{BA}$, then $A \bullet D \bullet B$.

2. Develop a proof for this statement:
   If m is a line and O, O^*, A are three points on m, then the halflines of m with respect to O relative to A are distinct from the halflines of m with respect to O^* relative to A.

3. Develop a proof for this statement:
   If m is a line and O, A, A^* are three points on m, then the halflines of m with respect to O relative to A are the same as the halflines of m with respect to O relative to A^*.

4.5 Segments; An Additional Axiom

A second special type of subset of a line is the topic presented in this section.

Definition 5: If A and B are two points, then the set of points $S = \{X \mid A \bullet X \bullet B\}$ is called a **segment** and is denoted as (A,B). The points A and B are called the **endpoints** of (A,B). (See **Figure 4.6**.)

Figure 4.6

Language Rule 9: The points A, B are said to **determine** (A,B).

Theorem 14: If A and B are two points, then $(A,B) = (B,A)$.

Theorem 15: If A and B are two points, then (A,B) is contained in line AB.

Theorem 16: If C is a point on a segment (A,B), then $\{C\}$, (A,C), and (C,B) partition (A,B).

Theorem 17: If C is a point on a segment (A,B), then $\{C\}$ separates (A,C) and (C,B).

While the following axiom played an integral role in Hilbert's geometry, it was first introduced in 1882 by the German mathematician M. Pasch.

Axiom 11: If A, B, C are three noncollinear points, m is a line which does not contain any of A, B, C, and m passes through a point of segment (A,B), then m also passes through a point of segment (B,C) or a point of segment (A,C). (See **Figure 4.7**.)

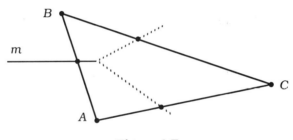

Figure 4.7

The following theorem is closely related to the preceding axiom in that it shows that the "or" used in the conclusion of the axiom's statement is an *exclusive* "or."

Theorem 18: If A, B, C are three noncollinear points and m is a line which passes through a point of segment (A,B) and a point of segment (B,C), then m cannot pass through a point of segment (A,C).

Proof: Suppose A, B, C are three noncollinear points and m is a line which passes, in turn, through a point D of segment (A,B), a point E of segment (B,C), and a point F of segment (A,C). (See **Figure 4.8**.)

By Axiom 1a, A, B, C determine lines AB, BC, AC, which are distinct, since A, B, C are noncollinear.

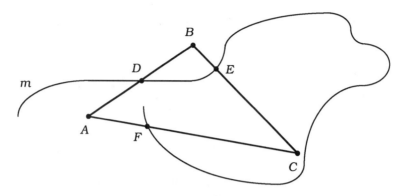

Figure 4.8

Since $D \in (A,B)$ and $F \in (A,C)$, then $A \bullet D \bullet B$ and $A \bullet F \bullet C$, respectively, by Definition 5. Thus, by Axiom 6, A and D are distinct and A and F are distinct. Also, by Theorem 15, $D \in AB$ and $F \in AC$, and, by Theorem 2, D and F are distinct, since AB and AC both contain A. Finally, A, D, F are noncollinear, by Axiom 1b, and, by Theorem 2, A, D, F cannot lie on BC.

Since $E \in (B,C)$, then, by Theorem 15, $E \in (BC)$.

By supposition, $D \bullet E \bullet F$, and hence, BC contains point E of (D,F).

Hence, A, D, F are distinct, noncollinear points, BC is a line which does not contain any of A, D, F, and BC contains E of (D,F). (See **Figure 4.9**.)

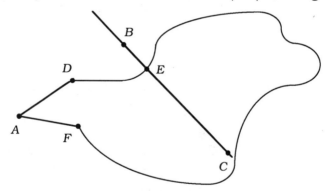

Figure 4.9

Claim: BC cannot contain a point of (A,D).

Proof: Suppose BC contains a point T of (A,D).

Since $(A,D) \subseteq AD = AB$, by Theorem 15, then, by Theorem 2, $T = B$, since AB and BC intersect at B.

Since $T \in (A,D)$, then $A \bullet T \bullet D$, by Definition 5. Thus, since $T = B$, then $A \bullet B \bullet D$, which contradicts Axiom 8, since $A \bullet D \bullet B$ already occurs.

Thus, BC cannot contain a point of (A,D). ◆

Claim: BC cannot contain a point of (A,F).

Proof: Analogous to the preceding **Claim**. ◆

Hence, Axiom 11 would be contradicted.

Thus, the supposition that m contains a point of (A,C) is false, and the theorem is true. ∎

Exercises 4.5

1. Develop a proof for Theorem 14.

2. Develop a proof for Theorem 15.

3. Develop a proof for Theorem 16.

4. Develop a proof for Theorem 17.

5. Develop two different proofs for this statement:
   If A and B are two points on a line m, then there exist at least three points which lie on m and are between A and B.
   (a) Use only axioms and theorems which appear in **Sections 4.2-4.3**.
   (b) Use any axioms and theorems, but use Axiom 5 to obtain only *one* point.

6. Develop a proof for this statement:
   If A, B, C are three noncollinear points and D, E are points such that $A \bullet B \bullet D$ and $B \bullet E \bullet C$, then line DE has a point in common with segment (A,C).

4.6 Halfplanes

The last type of plane subset to be considered in this chapter is that of halfplane. As has been the situation throughout this chapter, it will be assumed that all points lie on only one plane.

Definition 6: If m is a line on plane P and A is a point of P which is not on m, then
$$P_1 = \{A\} \cup \{X \mid X \in P, \ X \notin m, \text{ and } (A,X) \cap m = \emptyset\}$$
$$\text{and } P_2 = \{X \mid X \in P, \ X \notin m, \text{ and } (A,X) \cap m \neq \emptyset\}$$
are called the **halfplanes** of P with respect to m relative to A. (See **Figure 4.10**.)

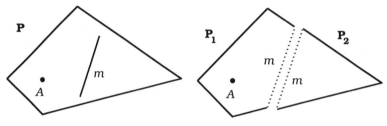

Figure 4.10

Language Rule 10: The halfplanes of plane P based on a line m can be described as having been **determined by m** and will be called the **sides of P relative to m** and **opposite halfplanes**. If A is a point in one of these halfplanes, then that halfplane will be called the **A-side of m**. Two points lying on the same halfplane will be described as being **on the same side of m**, while two points lying on different halfplanes will be described as being **on opposite sides of m**.

Language Rule 11: A **side of a segment** or a **side of a halfline** will refer to a side determined by the line which contains the segment or halfline.

Theorem 19: If m is a line on plane P and A is a point of P which is not on m, then m and P_1, P_2, the halfplanes of P with respect to m relative to A, partition P.

Theorem 20: If m is a line on plane P, then m separates the two halfplanes which it determines on P.

The following theorem concerning halflines can now be proved. This theorem will play a significant role in the development of proofs of numerous theorems in the remaining chapters.

Theorem 21: If a halfline has its endpoint on a line m but does not lie on *m*, then all the points of the halfline lie on the same side of *m*.

The terminology presented in Language Rule 10 can be used in the statement of the next theorem. A careful examination of the statement of this theorem and your proof of Theorem 20 should indicate that a proof of Theorem 22 would involve little more than a restatement of various parts of that proof.

Theorem 22: If *m* is a line and *A*, *B*, and *C* are three points not on *m*, then
(a) if *A* and *B* are on the same side of *m* and *B* and *C* are on the same side of *m*, then *A* and *C* are on the same side of *m*;
(b) if *A* and *B* are on the same side of *m* and *B* and *C* are on opposite sides of *m*, then *A* and *C* are on opposite sides of *m*;
(c) if *A* and *B* are on opposite sides of *m* and *B* and *C* are on opposite sides of *m*, then *A* and *C* are on the same side of *m*. (See **Figure 4.11**.)

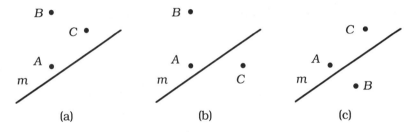

Figure 4.11

Exercises 4.6

1. Develop a proof for Theorem 19.

2. Develop a proof for Theorem 20.

3. Develop a proof for Theorem 21.

4. Develop a proof for this statement:
   If *m* and *m** are two lines on plane **P** and *A*, *A** are two points on **P** such that *A* is not on *m* and *A** is not on *m**, then the halfplanes of **P** with respect to *m* relative to *A* are distinct from the halfplanes of **P** with respect to *m** relative to *A**.

5. Develop a proof for this statement:
   If *m* is a line on plane ***P*** and *A*, *A** are two points on ***P*** which are not on *m*, then the halfplanes of ***P*** with respect to *m* relative to *A* are the same as the halfplanes of ***P*** with respect to *m* relative to *A**.

4.7 Some Additional Theorems Concerning Existence and Incidence

The collection of theorems in this section provides additional information concerning the existence of points and lines. These theorems will show that there are many more points and lines than have been used thus far in this development.

Theorem 23: There exist infinitely many distinct points on any segment.

Theorem 24: There exist infinitely many distinct points on any halfline.

Theorem 25: There exist infinitely many distinct points on any line.

Theorem 26: There exist infinitely many distinct lines through any point.

Theorem 27: If *A* is a point, then there exist infinitely many distinct lines which do not contain *A*.

Exercises 4.7

1. Develop a proof for Theorem 23.

2. Develop a proof for Theorem 24.

3. Develop a proof for Theorem 25.

4. Develop a proof for Theorem 26.

5. Develop a proof for Theorem 27.

4.8 Convexity

Although Hilbert did not mention the idea of convexity in his *Foundations*, it is a concept which blends well with an axiomatic development based on his axioms. As the theorems of this section will indicate, it is a concept which describes an interesting characteristic of some subsets of a plane. This material will be useful in establishing characteristics and theorems relating to additional types of plane subsets, which will be presented in the next chapter.

Definition 7: A set S of points is called **convex** if, for any two points A, B belonging to S, the segment (A,B) is a subset of S.

Theorem 28: Every line is convex.

Theorem 29: Each halfline of a line m, with respect to a point O on m, is convex.

Theorem 30: Every segment is convex.

Theorem 31: Each halfplane of a plane P, with respect to a line m on P, is convex.

The final theorem of this section is in no way related to Hilbert's axioms, but it will be of significance in developing proofs for some theorems in subsequent sections. Its proof is entirely dependent on a concept studied in set theory, namely the **Principle of Mathematical Induction**. For our purposes, the following is an appropriate form of this principle:

Principle of Mathematical Induction: If $\mathbf{S} = \{S_1, S_2, ..., S_k, ...\}$ is a collection of related statements such that
(a) S_1 is true; and
(b)   when S_k is assumed to be true, then S_{k+1} can be proved to be true,
then each statement in the collection $\mathbf{S}$ is true.

Theorem 32: The intersection of a finite number n, $n \geq 2$, of convex sets is convex.

Proof: Suppose $\mathbf{C}$ is a finite collection of convex sets.

Claim: If C_1 and C_2 are two members of $\mathbf{C}$, then $C_1 \cap C_2$ is convex.

Proof: Suppose each of C_1 and C_2 is a convex set, and suppose A and B are two points of $C_1 \cap C_2$.

Since A, B are in $C_1 \cap C_2$, then A, B are in C_1 and A, B are in C_2.

The segment (A,B) is in each of C_1 and C_2, since each of these sets is convex.

Hence, (A,B) is in $C_1 \cap C_2$.

Thus, $C_1 \cap C_2$ is convex. ◆

Claim: If C_1, C_2, ..., C_k, C_{k+1} are members of **C** and $\bigcap_{i=1}^{k} C_i$ is convex, then $\bigcap_{i=1}^{k+1} C_i$ is convex.

Proof: Suppose C_1, C_2, ..., C_k, C_{k+1} are members of **C** and $\bigcap_{i=1}^{k} C_i$ is convex.

$$\bigcap_{i=1}^{k+1} C_i = (C_1 \cap C_2 \cap ... \cap C_k \cap C_{k+1}) = \left(\bigcap_{i=1}^{k} C_i\right) \cap C_{k+1}$$

Each of C_{k+1} and $\bigcap_{i=1}^{k} C_i$ was assumed convex, so, by the preceding **Claim**, $\left(\bigcap_{i=1}^{k} C_i\right) \cap C_{k+1}$ is convex.

Thus, $\bigcap_{i=1}^{k+1} C_i$ is convex. ◆

Therefore, by the Principle of Mathematical Induction, the intersection of the members of **C** is convex, and the theorem is proved. ■

Exercises 4.8

1. Develop a proof for Theorem 28.

2. Develop a proof for Theorem 29.

3. Develop a proof for Theorem 30.

4. Develop a proof for Theorem 31.

SUPPLEMENTARY TOPICS

CONSTRUCTIONS

14. Using an initial segment measuring 1 inch, construct a segment whose length will be $3/5$ inch.

15. Using an initial segment measuring 1 inch, construct a segment whose length will be $\sqrt{3/5}$ inches.

16. Using an initial segment measuring 1 inch, construct a diagram containing a circle of radius 2 inches and a pair of parallel tangent lines to the circle. Determine all points which are equidistant from these two tangent lines and also exactly 1.5 inches from the circle.

17. Construct a triangle $\triangle ABC$ such that $\angle A$ is congruent to the given angle $\angle D$, side $\overline{AB}$ is congruent to the given segment $\overline{EF}$, and $m\overline{BC} + m\overline{AC} = m\overline{GH}$, which is given.

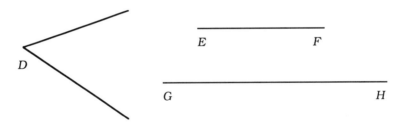

18. Using an initial segment measuring 1 inch, construct a segment whose length will be $\sqrt{3} + \sqrt{5}$ inches.

*****

PROBLEMS

8. The sides of a regular polygon of n, $n \geq 5$, sides are extended to form a star. Find an expression for the measure of the angle at each point of the star.

9. An equilateral triangle $\triangle AEF$ and a square $ABCD$ are positioned so that vertex A of $\triangle AEF$ lies on vertex A of $ABCD$ and triangle vertices E and F lie on square sides $\overline{BC}$ and $\overline{CD}$, respectively. If each side of the square has length 8 units, determine the area and perimeter of that portion of the square's interior which lies outside the triangle.

10. In the diagram below, $m\overline{BC} = 12$ units, $m\angle ABC = 30°$, $\overline{AB}$ is parallel to $\overline{CD}$, and O is the center of the circle. Determine the total area and total perimeter of the shaded regions.

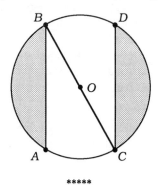

*****

TRANSFORMATION GEOMETRY

T-4: Rotations

Definition T-8: If C is a point in a plane, then a **rotation of $x°$ with center C** is a transformation of the plane such that

$$\mathbf{C}_x(P) = \begin{cases} P' = P, \text{ if } P = C \\ P', \text{ where } P' \text{ is the point on the circle} \\ \text{with radius } m\overline{CP} \text{ and center } C \text{ such} \\ \text{that } m\angle PCP' \text{ is } x°, \text{ if } P \neq C. \end{cases}$$

Remark T-2: A **rotation of $x°$ with center C** is often called an **$x°$ rotation about C**.

96 CHAPTER 4

Remark T-3: If $x > 0$, the rotation is in a counterclockwise direction. If $x < 0$, the rotation is in a clockwise direction. If $x = 0$, the rotation is called the **identity rotation**.

Definition T-9: A rotation in which $x = 180$ or $x = -180$ is called a **halfturn**.

Theorem T-6: A rotation is an isometry.

Example T-13: When a polygon is rotated $x°$ about a point C, its image can be found by rotating its vertices and sketching the polygon which they determine. (See **Figure T-29**.) ■

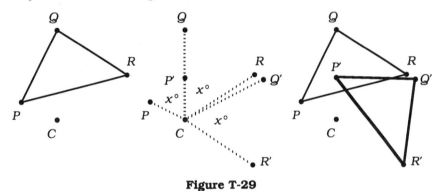

Figure T-29

Example T-14: When a line m is rotated $x°$ about a point C, its image can be found by rotating the point P, the foot of the perpendicular segment from C to m, and sketching the line which is perpendicular to the segment $\overline{CP'}$ at the point P'. (See **Figure T-30**.) ■

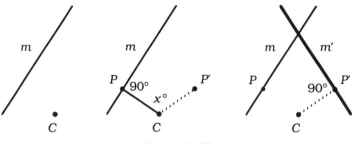

Figure T-30

Example T-15: If O is a circle and m is a line, as shown in **Figure T-31**, determine a segment $\overline{AB}$ such that $A \in m$, $B \in O$, and P is the midpoint

SUPPLEMENTARY TOPICS   97

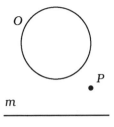

Figure T-31

of $\overline{AB}$.

Find $\mathbf{P}_{180}(O) = O'$, the image of O, by halfturning the center of O about P and using the radius of O to draw O'. (See **Figure T-32**.) Label a point where m and O' intersect as A. When A is halfturned about P, $\mathbf{P}_{180}(A)$ is the point of O which is the preimage of A; label $\mathbf{P}_{180}(A)$ as B. (See **Figure T-33**.)

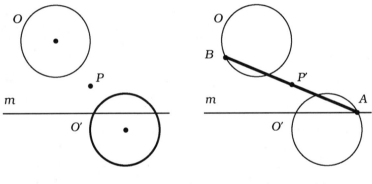

Figure T-32   **Figure T-33**

Segment $\overline{AB}$ is the desired segment. (*Note that a second such segment can be drawn, if the other point of $m \cap O'$ is selected to be A.*) ∎

Example T-16: If O is a circle and m is a line, as shown in **Figure T-34**, determine an equilateral triangle $\triangle PQR$ such the $Q \in m$ and $R \in O$.

Find $\mathbf{P}_{-60}(O)$ by rotating circle O $-60°$ about P. (See **Figure T-35**) Label a point where m and O' intersect as Q. $\mathbf{P}_{60}(Q)$ is the point of O which is the preimage of Q; label $\mathbf{P}_{60}(Q)$ as R. (See **Figure T-36**.)

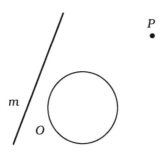

Figure T-34

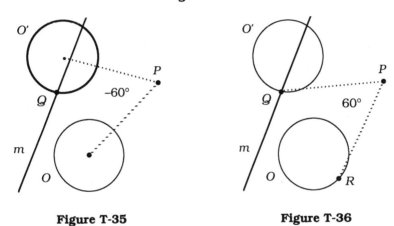

Figure T-35 **Figure T-36**

△PQR is the desired triangle. (See **Figure T-37**.) (*Note that a second such triangle can be drawn, if the other point of m ∩ O' is selected to be Q.*) ∎

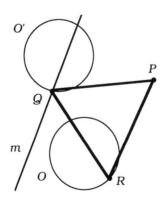

Figure T-37

Exercises T-4

1. Using a straightedge, draw a diagram similar to the one in **Figure T-38**. Use a rotation and constructions to determine a segment $\overline{DE}$ such that $D \in \triangle ABC$, $E \in m$, and P is the midpoint of $\overline{DE}$.

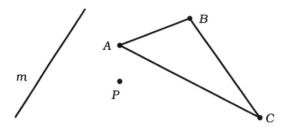

Figure T-38

2. Develop an informal proof for this statement:
   If C and A are two points in a plane and $\mathbf{C}_x(A) = B$, then C lies on the perpendicular bisector of the segment $\overline{AB}$.

3. Draw a diagram similar to that shown in **Figure T-39**. Use constructions to determine the center C of a rotation such that $\mathbf{C}_x(A) = A'$ and $\mathbf{C}_x(B) = B'$.

•B

•A'

•A B'•

Figure T-39

4. Does a rotation preserve or reverse orientation?

5. For the following statement, develop an informal proof for the special case which is illustrated in **Figure T-40**:
   If m and n are two lines which intersect at a point C and the angle of rotation from m to n is $x°$, then $\mathbf{R}_n(\mathbf{R}_m(P)) = \mathbf{C}_{2x}(P)$.

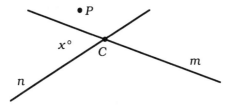

Figure T-40

6. Determine whether the following statement is true or false, and justify your answer:
   If a point P is reflected, successively, in two perpendicular lines which intersect at a point C, an equivalent result would be obtained by halfturning P about C.

7. Construct a diagram similar to that shown in **Figure T-41**. Use constructions to determine a square $ABCD$ such that $A \in m$ and $C \in O$.

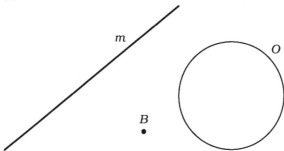

Figure T-41

8. Construct a collection of three parallel lines, with the middle one not being equidistant from the other two, and use constructions to determine an equilateral triangle which has one vertex on each of the three lines.

9. Construct two concentric circles, and label the smaller one as O_1 and the larger one as O_2. Select a point A which is inside O_2 and outside O_1. Construct an equilateral triangle $\triangle ABC$ such the $B \in O_1$ and $C \in O_2$. Is this always possible, regardless of the location of A?

10. Develop an informal proof for Theorem T-6.

Chapter 5: Triangles and Angles—Interiors and Exteriors

5.1 Introduction

The material in this chapter is a continuation of the study of subsets of a plane which was begun in the preceding chapter. Specifically, this chapter contains material pertaining to the interiors and exteriors of triangles and angles.

In the preceding chapter, Axiom 11 and Theorem 18 both implicitly pertained to the geometric concept of a triangle, although no definition was stated. Since this chapter contains a large number of theorems which describe characteristics of a triangle, it is now desirable to have such a definition.

Definition 8: If A, B, C are three noncollinear points, then the set which is the union of $\{A,B,C\}$ and the segments (A,B), (A,C), (B,C) is called a **triangle**. The points A, B, C are called **vertices** and (A,B), (A,C), (B,C) are called **sides**. The lines AB, AC, BC are called **sidelines**.

Definition 9: In a triangle, the vertex which is not an endpoint for a particular side is called the **opposite vertex** for that side, while the side is called the **opposite side** for that vertex.

Notation Rule 6: If A, B, C are the vertices of a triangle, then the triangle may be denoted as $\triangle ABC$.

5.2 Triangles—Interiors and Exteriors

Some interesting characteristics of triangles can be established with the use of some of the theorems of the preceding chapter and the following definitions.

Definition 10: The **interior** of a triangle $\triangle ABC$ is the intersection of
(a)   the side of line AB which contains C;
(b)   the side of line AC which contains B; and
(c)   the side of line BC which contains A. (See **Figure 5.1**)

Notation Rule 7: The interior of a triangle $\triangle ABC$ will be denoted as $int(\triangle ABC)$.

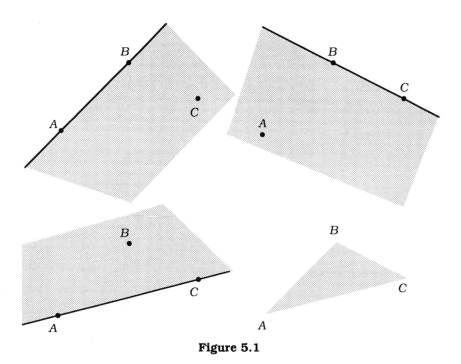

Figure 5.1

Definition 11: The **exterior** of a triangle $\triangle ABC$ is the set of all points which are neither on $\triangle ABC$ nor in $int(\triangle ABC)$. (See **Figure 5.2**)

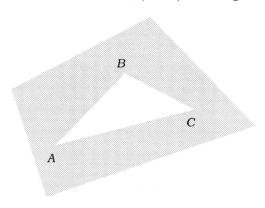

Figure 5.2

Notation Rule 8: The exterior of a triangle $\triangle ABC$ will be denoted as $ext(\triangle ABC)$.

Theorem 33: If D and E are points on two sides of a triangle $\triangle ABC$ and F is a point such that $D \bullet E \bullet F$, then F is in $int(\triangle ABC)$.

Proof: Suppose D and E are points on two sides of $\triangle ABC$ and F is a point such that $D \bullet E \bullet F$; without loss of generality, further suppose that $D \in (A,B)$ and $E \in (B,C)$. (See **Figure 5.3**.)

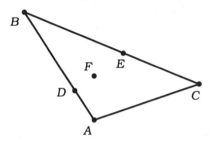

Figure 5.3

Since $D \in (A,B)$ and $E \in (B,C)$ then, by Definition 5, $A \bullet D \bullet B$ and $B \bullet E \bullet C$. By Axiom 6, A, D, B are distinct and collinear, as are B, E, C. By Theorem 15, $D \in AB$ and $E \in BC$. Hence, by Definition 2, $D \in \overrightarrow{AB}$ and $E \in \overrightarrow{BC}$. By Axiom 7 and Definition 2, $D \in \overrightarrow{BA}$ and $E \in \overrightarrow{CB}$.

Since $D \bullet F \bullet E$, then, by Axiom 6, D, F, E are collinear. Thus, $F \in DE$, and by, Definition 2, $F \in \overrightarrow{DE}$. By Axiom 7 and Definition 2, $F \in \overrightarrow{ED}$.

Claim: F is on the C-side of AB.

Proof: From the preceding statements, $D \in AB$, $E \in BC$, $E \in \overrightarrow{BC}$, and $F \in \overrightarrow{DE}$. (See **Figure 5.4**.)

Since A, B, C are noncollinear, from Definition 8, then AB and BC are distinct.

Since each of C and E is on BC and is different from B, then, by Theorem 2, neither of C and E is on AB.

Thus, each of $\overrightarrow{BC}$ and $\overrightarrow{DE}$ has its endpoint on AB but does not lie on AB. By Theorem 21, all points on $\overrightarrow{BC}$ lie on the C-side of AB and all points on $\overrightarrow{DE}$ lie on the E-side of AB.

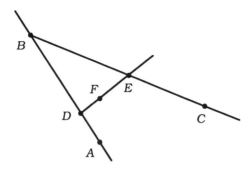

Figure 5.4

Hence, E is on the C-side of AB, and F is on the E-side of AB.
Thus, by Theorem 22, F is on the C-side of AB. ♦

Claim: F is on the A-side of BC.

Proof: Analogous to the preceding **Claim.** ♦

Claim: F is on the B-side of AC.

Proof: From the preceding statements, $D \in \overrightarrow{AB}$, $E \in \overrightarrow{CB}$, and $D \bullet F \bullet E$. (See **Figure 5.5**.)

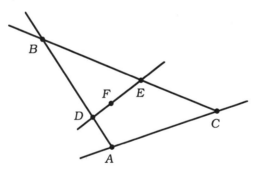

Figure 5.5

Since A, B, C are noncollinear, from Definition 8, then B is not on AC. Thus, each of $\overrightarrow{AB}$ and $\overrightarrow{CB}$ has its endpoint on AC but does not lie on AC. By Theorem 21, all points of $\overrightarrow{AB}$ and $\overrightarrow{CB}$ lie on the B-side of AC. Hence, each of D and E is on the B-side of AC.

By Theorem 31, each point of (D,E) lies on the B-side of AC. Since $D \bullet F \bullet E$, then, by Definition 5, $F \in (D,E)$.
Therefore, F is on the B-side of AC. ◆

Since F is on the C-side of AB, the A-side of BC, and the B-side of AC, then, by Definition 10, $F \in int(\triangle ABC)$. ∎

Theorem 34: The interior of a triangle is nonempty.

Theorem 35: If D and E are points on two sides of a triangle $\triangle ABC$ and F is a point on line DE and in $int(\triangle ABC)$, then $D \bullet F \bullet E$.

Theorem 36: If a line passes through a vertex of a triangle and contains a point in the triangle's interior, then it intersects the side opposite that vertex.

Exercises 5.2

1. Develop a proof for Theorem 34.

2. Develop a proof for Theorem 35.

3. Develop a proof for Theorem 36.

4. Develop a proof for this statement:
   The interior of a triangle is convex.

5. Develop a proof for this statement:
   A triangle is not convex.

6. Develop a proof for this statement:
   The exterior of a triangle is not convex.

7. Develop a proof for this statement:
   If a line contains a point in the interior of a triangle but does not pass through a vertex of the triangle, then it intersects at least one side of the triangle.

8. Develop a proof for this statement:
   If a line contains a point in the interior of a triangle and also inter-

sects only one side of the triangle, then it passes through the vertex which is opposite that side.

9. Develop a proof for this statement:
   If $\triangle ABC$ is on plane *P*, then $\triangle ABC$, $int(\triangle ABC)$, and $ext(\triangle ABC)$ partition *P*.

10. Develop a proof for this statement:
    If $\triangle ABC$ is on plane *P*, then $\triangle ABC$ does not separate $int(\triangle ABC)$ and $ext(\triangle ABC)$.

5.3 Angles—Interiors and Exteriors

As the definitions and theorems of this section will indicate, there is much similarity between characteristics of triangles and angles. In addition, some of the theorems pertaining to triangles will be used to prove theorems concerning angles.

Definition 12: If $\overrightarrow{OA}$ and $\overrightarrow{OB}$ are two noncollinear halflines which have *O* as a common endpoint, then the set which is the union of $\overrightarrow{OA}$, $\overrightarrow{OB}$, and $\{O\}$ is called an **angle**. The point *O* is called the **vertex** of the angle. The halflines $\overrightarrow{OA}$ and $\overrightarrow{OB}$ are called **sides** of the angle, and the lines *OA* and *OB* are called **sidelines**.

Notation Rule 9: The angle which is the union of the halflines $\overrightarrow{OA}$ and $\overrightarrow{OB}$ and the point *O* may be denoted as $\angle AOB$. When *O* serves as the vertex of only one angle, that angle may be denoted as $\angle O$.

Notation Rule 10: If *A* and *B* are any two points, with one located on each of the sides of an angle $\angle O$, then the angle may be denoted as $\angle AOB$.

Language Rule 12: In a triangle, **angle of the triangle** will refer to an angle which has for its vertex a vertex of the triangle and whose sides contain the sides of the triangle for which the vertex is a common endpoint. **Angle opposite a side** will refer to the angle of the triangle which has for its vertex the triangle vertex that is opposite the particular side. **Side op-**

posite an angle will refer to the side of the triangle which is opposite the triangle vertex that is the vertex of the particular angle. An angle will be referred to as being **included between two sides** when its sides contain those sides of the triangle. A side will be referred to as being **included between two angles** when its endpoints are the vertices of those angles of the triangle.

Definition 13: The **interior** of an angle $\angle AOB$ is the intersection of

(a) the side of halfline $\overrightarrow{OA}$ which contains B; and

(b) the side of halfline $\overrightarrow{OB}$ which contains A. (See **Figure 5.6**.)

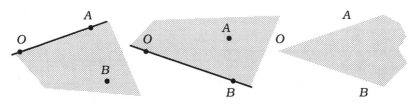

Figure 5.6

Notation Rule 11: The interior of an angle $\angle AOB$ will be denoted as $int(\angle AOB)$.

Definition 14: The **exterior** of an angle $\angle AOB$ is the set of all points which are neither on $\angle AOB$ nor in $int(\angle AOB)$. (See **Figure 5.7**.)

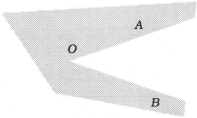

Figure 5.7

Notation Rule 12: The exterior of an angle $\angle AOB$ will be denoted as $ext(\angle AOB)$.

Theorem 37: The interior of an angle is nonempty.

Theorem 38: If A and B are points on two sides of an angle $\angle O$ and C

is a point such that $A \bullet C \bullet B$, then C is in $int(\angle O)$.

Theorem 39: If A and B are points on two sides of an angle $\angle O$ and C is a point on line AB and in $int(\angle O)$, then $A \bullet C \bullet B$.

Theorem 40: If a point C lies in the interior of an angle $\angle AOB$, then the halfline $\overrightarrow{OC}$ is a subset of $int(\angle AOB)$.

Theorem 41: If point C lies in the interior of an angle $\angle AOB$, then halfline $\overrightarrow{OC}$ intersects the segment (A,B).

Theorem 42: If a point C lies in the interior of an angle $\angle AOB$ and a point D lies in the exterior of $\angle AOB$, then the segment (C,D) intersects $\angle AOB$.

Proof: Suppose C is a point which lies in the interior of an angle $\angle AOB$ and D is a point which lies in the exterior of $\angle AOB$.

Since D lies in $ext(\angle AOB)$, then D is not on $\angle AOB$ and D is not in $int(\angle AOB)$. By Definition 13, D and B are on opposite sides of OA and/or D and A are on opposite sides of OB. Hence, there are five possible locations for D:

(i) D is on the halfline of OA which does not contain A; that is, D and A are on opposite sides of O, and $\overrightarrow{OD}$ and $\overrightarrow{OA}$ are the two halflines of OA relative to O.

(ii) D is on the halfline of OB which does not contain B; that is, D and B are on opposite sides of O, and $\overrightarrow{OD}$ and $\overrightarrow{OB}$ are the two halflines of OB relative to O.

(iii) D and A are on opposite sides of OB, and D and B are on the same side of OA.

(iv) D and B are on opposite sides of OA, and D and A are on the same side of OB.

(v) D and A are on opposite sides of OB, and D and B are on opposite sides of OA. (See **Figure 5.8**.)

Claim: If D, A are on opposite sides of O, then (C,D) intersects $\angle AOB$.

Proof: Suppose C lies in $int(\angle AOB)$ and D and A are on opposite sides

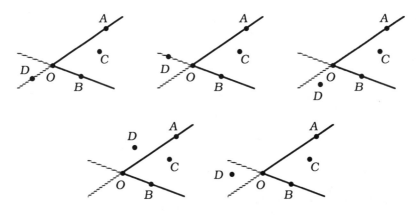

Figure 5.8

of O.

Since D and A are on opposite sides of O, then, by Definition 2, $A \bullet O \bullet D$. Thus, since $O \in OB$, then A and D are on opposite sides of OB.

Since $C \in int(\angle AOB)$, C is on the A-side of OB and C is on the B-side of OA.

Thus, by Theorem 22, C and D are on opposite sides of OB. Hence, by Theorem 20 and Definition 4, there exists a point E on OB such that $C \bullet E \bullet D$. By Definition 5, $E \in (C,D)$.

There are two possibilities: $E = B$ or $E \neq B$.

Subclaim: If $E = B$, then (C,D) intersects $\angle AOB$.

Proof: Suppose $E = B$.
Since B is on $\angle AOB$, then (C,D) intersects $\angle AOB$. ◆◆

Subclaim: If $E \neq B$, then (C,D) intersects $\angle AOB$.

Proof: Suppose $E \neq B$

Since $C \bullet E \bullet D$, then, by Axiom 7 and Definition 2, $E \in \overrightarrow{DC}$. Since D is on OA and C is on the B-side of OA, then, by Theorem 21, all the points of $\overrightarrow{DC}$ lie on the B-side of OA. Thus, E is on the B-side of OA.

Since E is on the B-side of OA and $E \in OB$, then $B \bullet O \bullet E$ is not possible. Thus, by Definition 2, $E \in \overrightarrow{OB}$.

Hence, (C,D) intersects $\angle AOB$. ♦♦

Therefore, if D, A are on opposite sides of O, then (C,D) intersects $\angle AOB$. ♦

Claim: If D, B are on opposite sides of O, then (C,D) intersects $\angle AOB$.

Proof: Analogous to the preceding **Claim**. ♦

Claim: If D, A are on opposite sides of OB and D, B are on the same side of OA, then (C,D) intersects $\angle AOB$.

Proof: Suppose $C \in int(\angle AOB)$, D and A are on opposite sides of OB, and D and B are on the same side of OA.
Since $C \in int(\angle AOB)$, then C is on the A-side of OB. Hence, by Theorem 22, D and C are on opposite sides of OB.
By Theorem 20 and Definition 4, there exists a point E on OB such that $C \bullet E \bullet D$. By Definition 5, $E \in (C,D)$.
There are two possibilities: $E = B$ or $E \neq B$.

Subclaim: If $E = B$, then (C,D) intersects $\angle AOB$.

Proof: Suppose $E = B$.
Since B is on $\angle AOB$, then (C,D) intersects $\angle AOB$. ♦♦

Subclaim: If $E \neq B$, then (C,D) intersects $\angle AOB$.

Proof: Suppose $E \neq B$.
Since each of C and D is on the B-side of OA, then, by Theorem 31, (C,D) is a subset of the B-side of OA. Thus, E is on the B-side of OA.
Since E is on the B-side of OA and $E \in OB$, then $B \bullet O \bullet E$ is not possible. Thus, by Definition 2, $E \in \overrightarrow{OB}$.
Hence, (C,D) intersects $\angle AOB$. ♦♦

Therefore, if D, A are on opposite sides of OB and D, B are on the same side of OA, then (C,D) intersects $\angle AOB$. ♦

Claim: If D, B are on opposite sides of OA and D, A are on the same side of OB, then (C,D) intersects $\angle AOB$.

Proof: Analogous to the preceding **Claim**. ◆

Claim: If D, A are on opposite sides of OB and D, B are on opposite sides of OA, then (C,D) intersects $\angle AOB$.

Proof: Suppose C is in $int(\angle AOB)$, D, A are on opposite sides of OB, and D, B are on opposite sides of OA.

Since $C \in int(\angle AOB)$, C is on the B-side of OA and the A-side of OB. Hence, by Theorem 22, C, D are on opposite sides of OA, and C, D are on opposite sides of OB.

By Theorem 20 and Definition 4, there exists a point E on OA such that $C \bullet E \bullet D$. By Definition 5, $E \in (C,D)$.

There are two possibilities: $E = A$ or $E \neq A$.

Subclaim: If $E = A$, then (C,D) intersects $\angle AOB$.

Proof: Suppose $E = A$.
Since A is on $\angle AOB$, then (C,D) intersects $\angle AOB$. ◆◆

Subclaim: If $E \neq A$, then (C,D) intersects $\angle AOB$.

Proof: Suppose $E \neq A$.

There are two possibilities: $E \in \overrightarrow{OA}$ or E is on the opposite halfline of $\overrightarrow{OA}$ relative to O; denote this halfline as $\overrightarrow{OZ}$.

If $E \in \overrightarrow{OA}$, then (C,D) intersects $\angle AOB$.

Now suppose $E \in \overrightarrow{OZ}$.

With an argument analogous to that used in the first **Claim** it can be shown that (C,E) intersects $\angle AOB$ at a point F. By Definition 6, $C \bullet F \bullet E$.

Since $C \bullet E \bullet D$ and $C \bullet F \bullet E$, then, by Theorem 7 and Axiom 10, $C \bullet F \bullet D$. By Definition 6, $F \in (C,D)$.

Hence, (C,D) intersects $\angle AOB$. ◆◆

Therefore, if D, A are on opposite sides of OB and D, B are on opposite sides of OA, then (C,D) intersects ∠AOB. ♦

Thus, if a point C lies in the interior of an angle ∠AOB and a point D lies in the exterior of ∠AOB, then the segment (C,D) intersects ∠AOB. ∎

Exercises 5.3

1. Develop a proof for Theorem 37.

2. Develop a proof for Theorem 38.

3. Develop a proof for Theorem 39.

4. Develop a proof for Theorem 40.

5. Develop a proof for Theorem 41.

6. Develop a proof for this statement:
   The interior of an angle is convex.

7. Develop a proof for this statement:
   An angle is not convex.

8. Develop a proof for this statement:
   The exterior of an angle is not convex.

9. Develop a proof for this statement:
   If ∠AOB is on plane **P**, then ∠AOB, int(AOB), and ext(∠AOB) partition **P**.

10. Why can Theorem 42 not be replaced with the following statement?
    If ∠AOB is on plane **P**, then ∠AOB separates int(AOB) and ext(∠AOB).

5.4 Betweenness for Halflines

Just as there was a betweenness characteristic for collinear points, as presented in Section 4.3, there is a betweenness property for halflines which have the same endpoint. As the following definition indicates, this property for halflines is expressed with the use of the concept of angle interior.

Definition 15: If halfline $\overrightarrow{OC}$ is a subset of the interior of $\angle AOB$, then $\overrightarrow{OC}$ is said to be **between** $\overrightarrow{OA}$ and $\overrightarrow{OB}$.

This definition can be used with some of the preceding theorems to establish some theorems relating to the betweenness property for halflines.

Theorem 43: If $\overrightarrow{OA}$ and $\overrightarrow{OB}$ are noncollinear halflines, then a third halfline $\overrightarrow{OC}$ is between $\overrightarrow{OA}$ and $\overrightarrow{OB}$ if and only if there exist points A^*, C^*, B^* on $\overrightarrow{OA}$, $\overrightarrow{OC}$, $\overrightarrow{OB}$, respectively, such that $A^* \bullet C^* \bullet B^*$.

Theorem 44: If $\overrightarrow{OA}$, $\overrightarrow{OB}$, $\overrightarrow{OC}$, $\overrightarrow{OD}$ are halflines, with $\overrightarrow{OA}$ and $\overrightarrow{OD}$ being opposite halflines, and $\overrightarrow{OB}$ is between $\overrightarrow{OA}$ and $\overrightarrow{OC}$, then $\overrightarrow{OC}$ is between $\overrightarrow{OB}$ and $\overrightarrow{OD}$.

Theorem 45: If $\overrightarrow{OA}$, $\overrightarrow{OB}$, $\overrightarrow{OC}$, $\overrightarrow{OD}$ are halflines, with $\overrightarrow{OA}$, $\overrightarrow{OB}$ being noncollinear, such that $\overrightarrow{OC}$ is between $\overrightarrow{OA}$, $\overrightarrow{OB}$ and $\overrightarrow{OD}$ is between $\overrightarrow{OB}$, $\overrightarrow{OC}$, then $\overrightarrow{OD}$ is between $\overrightarrow{OA}$ and $\overrightarrow{OB}$.

Theorem 46: If $\overrightarrow{OA}$, $\overrightarrow{OB}$, $\overrightarrow{OC}$, $\overrightarrow{OD}$ are halflines, with $\overrightarrow{OA}$, $\overrightarrow{OB}$ noncollinear, $\overrightarrow{OC}$, $\overrightarrow{OD}$ noncollinear, D on the C-side of $\overrightarrow{OA}$, such that $\overrightarrow{OC}$ is between $\overrightarrow{OA}$, $\overrightarrow{OB}$ and $\overrightarrow{OB}$ is between $\overrightarrow{OC}$, $\overrightarrow{OD}$, then $\overrightarrow{OB}$ is between $\overrightarrow{OA}$ and $\overrightarrow{OD}$.

Exercises 5.4

1. Develop a proof for Theorem 43.

2. Develop a proof for Theorem 44.

3. Develop a proof for Theorem 45.

4. Develop a proof for Theorem 46.

****

SUPPLEMENTARY TOPICS

CONSTRUCTIONS

19. Using the two given segments, whose measures are p and q units, respectively, construct a segment whose measure will be $\sqrt{3pq}$ units.

20. Using the two given segments, whose measures are c and d units, respectively, construct a segment whose measure will be c^2/d.

21. Select any three noncollinear points A, B, C. Determine all points which are equidistant from these points.

22. Construct a triangle $\triangle ABC$ such that $\angle A$ is congruent to the given angle $\angle D$, $\angle B$ is congruent to the given angle $\angle E$, and $m\overline{AB} + m\overline{BC} + m\overline{AC} = m\overline{FG}$, which is given.

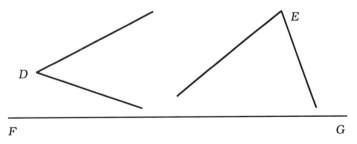

23. Using a straightedge, draw a diagram similar to the one given below, in which portions of both sides of an angle are shown. Without using the portion of the angle which is not given, construct the bisector of the angle.

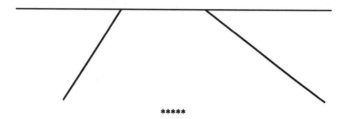

*****

PROBLEMS

11. Two diagonals of a regular pentagon intersect. Show that the measure of the longer segment of each diagonal is equal to the measure of a side of the pentagon.

12. In the diagram below, each side of square $ABCD$ has measure 4 units, $\overline{CF} \cong \overline{DF}$, $\overline{AE} \cong \overline{DE}$, $\angle DAE \cong \angle CDF$, and $m\angle CDF = 15°$. Determine the area and perimeter of the region enclosed by the quadrilateral $ABCF$.

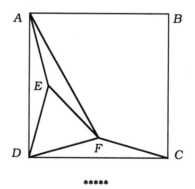

*****

TRANSFORMATION GEOMETRY

T-5: Translations

Definition T-10: A **translation** is a transformation of the plane such that each point of the plane is moved the same distance in the same direction.

Remark T-4: A **translation** is often called a **glide**.

Remark T-5: If a translation involves a zero distance, it is called the **identity translation**.

Theorem T-7: A translation is an isometry.

Example T-17: A translation **T** maps the origin to the point $(2,5)$. Determine the image of the point $(-3,4)$ and the preimage of the point $(5,-1)$.
$\mathbf{T}(0,0) = (0+2, 0+5) = (2,5)$.
Hence, $\mathbf{T}(-3,4) = (-3+2, 4+5) = (-1,9)$, and the image of $(-3,4)$ is $(-1,9)$.
If $\mathbf{T}(x,y) = (x+2, y+5) = (5,-1)$, then $(x,y) = (3,-6)$, and the preimage of $(5,-1)$ is $(3,-6)$. Thus, $\mathbf{T}(3,-6) = (5,-1)$. (See **Figure T-42**.)

The segments $\overline{(0,0)(2,5)}$, $\overline{(-3,4)(-1,9)}$, and $\overline{(3,-6)(5,-1)}$ all have length $\sqrt{29}$ units. In addition, all three segments have slope $5/2$. ∎

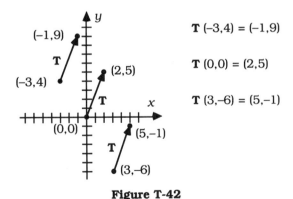

Figure T-42

Example T-18: When the image, under a translation **T**, of one vertex of a polygon is known, the image of the entire polygon can be determined by translating all the other vertices and sketching the polygon which has the images for its vertices. (See **Figure T-43**.) ∎

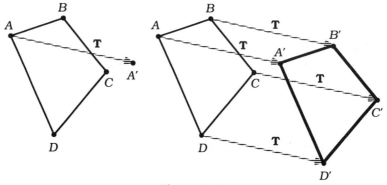

Figure T-43

Example T-19: If C_1 and C_2 are two circles, as shown in **Figure T-44**, determine segment $\overline{AB}$ such that $A \in C_1$, $B \in C_2$, and $\overline{AB}$ is congruent and parallel to the given segment $\overline{PP'}$, where $P' = \mathbf{T}(P)$.

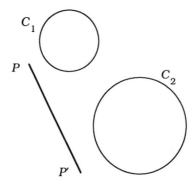

Figure T-44

Find $\mathbf{T}(C_1) = C_1'$, the image of C_1, by translating the center of C_1 and using the radius of C_1 to draw C_1'. (See **Figure T-45**.) Label a point where C_1' and C_2 intersect as B. The preimage of B will be the desired point A. (See **Figure T-46**.)

Segment $\overline{AB}$ is the desired segment. (Note that a second such segment can be drawn, if the other point of $C_1' \cap C_2$ is selected to be B.) ∎

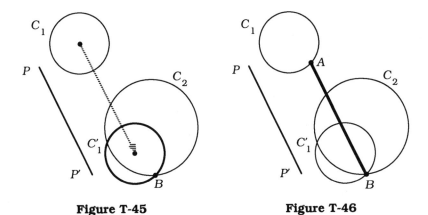

Figure T-45 **Figure T-46**

Definition T-11: A **glide reflection G** is a transformation of the plane involving a translation **T** and a line m parallel to the translation such that $G(P) = P'$, where $T(P) = P*$ and $R_m(P*) = P'$.

Theorem T-8: A glide reflection is an isometry.

Example T-20: A glide reflection **G** having the graph of $y = 3$ for its reflecting line maps the point $(2,7)$ to the point $(-6,-1)$. Determine the image of the point $(-1,-4)$ and the preimage of the point $(1,6)$.

For **G**, $T(2,7) = (2-8, 7+0) = (-6,7)$, and $R_{y=3}(-6,7) = (-6,-1)$.

Hence, $T(-1,-4) = (-1-8, -4+0) = (-9,-4)$ and $R_{y=3}(-9,-4) = (-9,10)$. Thus, $G(-1,-4) = (-9,10)$.

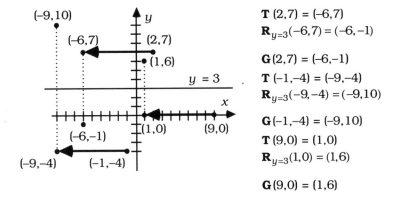

$T(2,7) = (-6,7)$
$R_{y=3}(-6,7) = (-6,-1)$

$G(2,7) = (-6,-1)$
$T(-1,-4) = (-9,-4)$
$R_{y=3}(-9,-4) = (-9,10)$

$G(-1,-4) = (-9,10)$
$T(9,0) = (1,0)$
$R_{y=3}(1,0) = (1,6)$

$G(9,0) = (1,6)$

Figure T-47

SUPPLEMENTARY TOPICS

If $\mathbf{R}_{y=3}(x,y) = (1,6)$, then $(x,y) = (1,0)$. Thus, $\mathbf{R}_{y=3}(1,0) = (1,6)$. If $\mathbf{T}(u,v) = (u-8, v+0) = (1,0)$, then $(u,v) = (9,0)$. Thus, the preimage of $(1,6)$ is $(9,0)$; that is, $\mathbf{G}(9,0) = (1,6)$. (See **Figure T-47**.) ∎

Example T-21: When the image, under a glide reflection **G**, of one vertex of a polygon is known, as shown in **Figure T-48**, the image of the entire polygon can be determined by glide reflecting all the other vertices and sketching the polygon which has the images as its vertices, as shown in **Figure T-49**. ∎

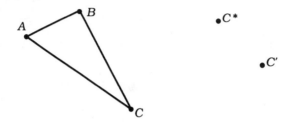

Figure T-48

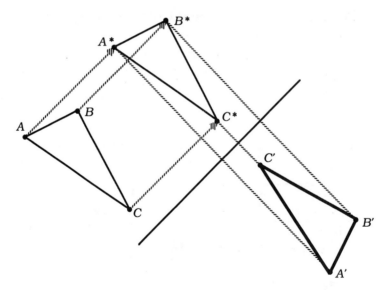

Figure T-49

Exercises T-5

1. A translation maps $(-2,5)$ to $(3,-1)$. Determine the image of $(4,3)$ and the preimage of $(0,-5)$.

2. Using a straightedge, draw a diagram similar to the one in **Figure T-50**. Use constructions to determine a segment $\overline{AB}$ such that $A \in m$, $B \in n$, and $\overline{AB}$ is congruent and parallel to $\overline{PP'}$, where $P' = \mathbf{T}(P)$.

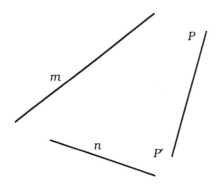

Figure T-50

3. Using a straightedge, draw a diagram similar to the one in **Figure T-51**. Assume the shaded portion of the diagram represents a river, and construct a minimum-distance line-segment path from A to B which crosses the river in a segment perpendicular to the parallel river banks. How many such paths are possible?

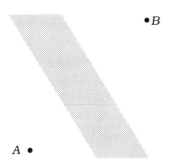

Figure T-51

4. Develop an informal proof for this statement:
   If **T** is a translation and A, B, $A' = \mathbf{T}(A)$, $B' = \mathbf{T}(B)$ are four noncollinear points, then the quadrilateral $AA'B'B$ is a parallelogram.

5. A glide reflection **G** having the graph of $y = -x$ for its reflecting line maps the point $(-1,5)$ to $(-1,-3)$. Determine the images of $(-2,0)$ and $(4,-4)$. Determine the preimage of $(-3,1)$.

6. Develop an informal proof for this statement:
   If **G** is a glide reflection involving the reflecting line m and P is any point of the plane not on m, then m bisects the segment $\overline{PP'}$, where $P' = \mathbf{G}(P)$.

7. Using a straightedge, draw a diagram similar to the one in **Figure T-52**. Use the statement in problem 6 to construct the reflecting line of the glide reflection which maps $\triangle ABC$ to $\triangle A'B'C'$. Finally, construct the translation image $\triangle A^*B^*C^*$ for $\triangle ABC$.

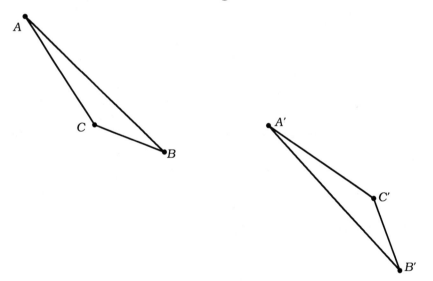

Figure T-52

8. Determine whether the following statement is true or false, and justify your answer:
   If a point P is reflected, successively, in parallel lines m and n

which are *k* units apart, an equivalent result would be obtained by translating *P* a distance of 2*k* units in the direction from *m* to *n*.

9. What is the relationship between translations and vectors?

10. Does a translation preserve or reverse orientation?

11. Does a glide reflection preserve or reverse orientation?

12. Explain why the translating and reflecting transformations of a glide reflection can be interchanged.

13. Develop an informal proof for Theorem T-7.

14. Develop an informal proof for Theorem T-8.

Chapter 6: Congruences and Comparisons

6.1 Introduction

The study of the geometry of a plane is continued with a study of the ideas of congruence and comparison. Congruence will be examined relative to segments, angles, and triangles, while comparison will be examined relative to segments and angles.

As was the case with point, line, and plane, the idea of congruence cannot be defined.

Language Rule 13: Congruence will be an undefined term.

Notation Rule 13: The symbol "≅" will be used to represent **congruent**.

6.2 Segment Congruences

Since undefined terms are characterized with axioms, it is necessary to begin this section with some axioms pertaining to segments and congruence. These axioms can be used in developing the proofs of two theorems pertaining to segments.

Axiom 12: If A and B are two points on a line m and C is a point on a line n, then there exist on n exactly two points D and E, one on each side of C, such that $(A,B) \cong (C,D)$ and $(A,B) \cong (C,E)$. (See **Figure 6.1**.)

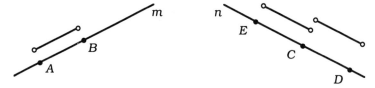

Figure 6.1

Axiom 13: (a) If (A,B) is a segment, then $(A,B) \cong (A,B)$.
(b) If (A,B) and (C,D) are segments such that $(A,B) \cong (C,D)$, then $(C,D) \cong (A,B)$.
(c) If (A,B), (C,D), (E,F) are segments such that $(A,B) \cong (C,D)$ and $(C,D) \cong (E,F)$, then $(A,B) \cong (E,F)$.

Axiom 14: If A, B, C, D, E, F are points such that $A \bullet B \bullet C$, $D \bullet E \bullet F$, $(A,B) \cong (D,E)$, and $(B,C) \cong (E,F)$, then $(A,C) \cong (D,F)$. (See **Figure 6.2**.)

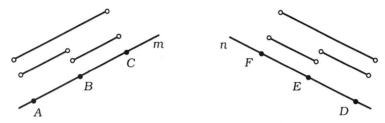

Figure 6.2

Theorem 47: If A, B, C, D, E, F are points such that $A \bullet B \bullet C$, $D \bullet E \bullet F$, $(A,B) \cong (D,E)$, and $(A,C) \cong (D,F)$, then $(B,C) \cong (E,F)$.

Proof: Suppose A, B, C, D, E, F are points such that $A \bullet B \bullet C$, $D \bullet E \bullet F$, $(A,B) \cong (D,E)$, $(A,C) \cong (D,F)$, and $(B,C) \not\cong (E,F)$.

There exists on $\overrightarrow{EF}$, by Axiom 12, a point G such that $(B,C) \cong (E,G)$.

Since $(B,C) \not\cong (E,F)$, $(B,C) \cong (E,G)$, and $G \in \overrightarrow{EF}$, then, by Axiom 12, $F \neq G$.

Claim: $D \bullet E \bullet G$.

Proof: Since $G \in \overrightarrow{EF}$, then, by Definition 2, $E \bullet G \bullet F$, $G = F$, or $E \bullet F \bullet G$.
If $E \bullet G \bullet F$, and $D \bullet E \bullet F$, then, by Theorem 8, $D \bullet E \bullet G \bullet F$. By Axiom 9, $D \bullet E \bullet G$.
If $G = F$, and $D \bullet E \bullet F$, then $D \bullet E \bullet G$.
If $E \bullet F \bullet G$, and $D \bullet E \bullet F$, then, by Theorem 6, $D \bullet E \bullet F \bullet G$. By Axiom 9, $D \bullet E \bullet G$.

Thus, if $G \in \overrightarrow{EF}$, then $D \bullet E \bullet G$. ♦

Since A, B, C, D, E, G are points such that $A \bullet B \bullet C$, $D \bullet E \bullet G$, $(A,B) \cong (D,E)$, and $(B,C) \cong (E,G)$, then, by Axiom 14, $(A,C) \cong (D,G)$.

Hence, $(A,C) \cong (D,F)$ and $(A,C) \cong (D,G)$. By Axiom 13, $(D,F) \cong (D,G)$.

Since $D \bullet E \bullet F$ and $D \bullet E \bullet G$, then, by Definition 2, F and G are on the same halfline determined by D and containing E. Hence, by Axiom 12, $F = G$.

Therefore, $F \neq G$ and $F = G$. Thus, the supposition $(B,C) \not\cong (E,F)$ is false.

Hence, $(B,C) \cong (E,F)$. ∎

Theorem 48: If A, B, C, D, E are points such that $A \bullet B \bullet C$ and $(A,C) \cong (D,E)$, then there exists exactly one point F such that $(A,B) \cong (D,F)$ and $D \bullet F \bullet E$.

Exercises 6.2

1. Develop a proof for Theorem 48.

2. Show that congruence is an equivalence relation on the set of all segments on a plane.

6.3 Angle and Triangle Congruences

It is necessary that some axioms be stated concerning angles and the undefined ideas of congruence. With the use of an additional axiom relating triangles, angles, segments, and the congruence concept, it will be possible to prove several theorems concerning congruent angles and congruent triangles.

Axiom 15: If $\angle ABC$ is an angle and $\overrightarrow{DE}$ is a halfline on a line m, then there exist exactly two halflines $\overrightarrow{DF}$ and $\overrightarrow{DG}$, one on each side of m, such that $\angle ABC \cong \angle EDF$ and $\angle ABC \cong \angle EDG$. (See **Figure 6.3**.)

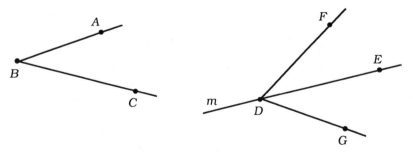

Figure 6.3

Axiom 16: (a) If ∠ABC is an angle, then ∠ABC ≅ ∠ABC.
(b) If ∠ABC and ∠DEF are angles such ∠ABC ≅ ∠DEF, then ∠DEF ≅ ∠ABC.
(c) If ∠ABC, ∠DEF, ∠GHI are angles such that ∠ABC ≅ ∠DEF and ∠DEF ≅ ∠GHI, then ∠ABC ≅ ∠GHI.

Definition 16: If △ABC and △DEF are triangles such that a one-to-one correspondence can be established between their vertices such that corresponding sides are congruent and corresponding angles are congruent, then △ABC and △DEF are called **congruent triangles**.

Notation Rule 14: If △ABC ↔ △DEF, then the following correspondences occur among their respective parts:

$$A \leftrightarrow D \qquad \angle A \leftrightarrow \angle D \qquad (A,B) \leftrightarrow (D,E)$$
$$B \leftrightarrow E \qquad \angle B \leftrightarrow \angle E \qquad (B,C) \leftrightarrow (E,F)$$
$$C \leftrightarrow F \qquad \angle C \leftrightarrow \angle F \qquad (A,C) \leftrightarrow (D,F)$$

Theorem 49: (a) If △ABC is a triangle, then △ABC ≅ △ABC.
(b) If △ABC and △DEF are triangles such that △ABC ≅ △DEF, then △DEF ≅ △ABC.
(c) If △ABC, △DEF, △GHI are triangles such that △ABC ≅ △DEF and △DEF ≅ △GHI, then △ABC ≅ △GHI.

Axiom 17: If △ABC and △DEF are triangles such that $(A,B) \cong (D,E)$, ∠B ≅ ∠E, and $(B,C) \cong (E,F)$, then ∠A ≅ ∠D and ∠C ≅ ∠F. (See **Figure 6.4**.)

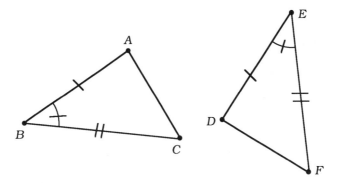

Figure 6.4

CONGRUENCES AND COMPARISONS

Theorem 50: If $\triangle ABC$ and $\triangle DEF$ are triangles such that $(A,B) \cong (D,E)$, $\angle B \cong \angle E$, and $(B,C) \cong (E,F)$, then $\triangle ABC \cong \triangle DEF$.

Proof: Suppose $\triangle ABC$ and $\triangle DEF$ are triangles such that $(A,B) \cong (D,E)$, $\angle B \cong \angle DEF$, and $(B,C) \cong (E,F)$. (See **Figure 6.5**.)

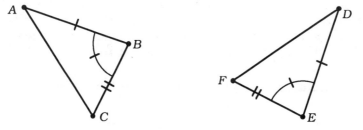

Figure 6.5

By Axiom 17, $\angle A \cong \angle D$ and $\angle C \cong \angle F$.

Claim: $(A,C) \cong (D,F)$.

Proof: By Theorem 15, (D,F) lies on line DF. By Axiom 12, there exists exactly one point G lying on $\overrightarrow{DF}$ for which $(A,C) \cong (D,G)$. (See **Figure 6.6**.) Thus, in $\triangle ABC$ and $\triangle DEG$, $(A,B) \cong (D,E)$, $\angle A \cong \angle D$, and $(A,C) \cong (D,G)$.

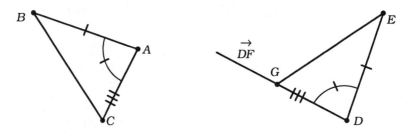

Figure 6.6

Hence, by Axiom 17, $\angle B \cong \angle DEG$ and $\angle C \cong \angle G$.

Therefore, by Axiom 16, $\angle DEF \cong \angle DEG$, since $\angle B \cong \angle DEF$ and $\angle B \cong \angle DEG$.

By Axiom 15, $\overrightarrow{EF} \cong \overrightarrow{EG}$. Hence, $EF = EG$, and, by Theorem 2, $F = G$. Thus, $(A,C) \cong (D,F)$. ◆

Therefore, in $\triangle ABC$ and $\triangle DEF$, $(A,B) \cong (D,E)$, $\angle B \cong \angle E$, $(B,C) \cong (E,F)$, $\angle A \cong \angle D$, $\angle C \cong \angle F$, and $(A,C) \cong (D,F)$.

By Definition 16, $\triangle ABC \cong \triangle DEF$. ∎

Language Rule 14: The preceding theorem can be called the **Side Angle Side** congruence theorem and restated as "If two sides and the included angle of one triangle are congruent, respectively, to two sides and the included angle of another triangle, then the triangles are congruent."

An interesting use for this theorem concerning triangle congruence is in establishing a characteristic of a special type of triangle, which is described in the next definition. The theorem concerning this special type of triangle will be used to establish some later theorems.

Definition 17: A triangle is called **isosceles** if two of its sides are congruent.

Theorem 51: If $\triangle ABC$ is a triangle such that $(A,B) \cong (A,C)$, then $\angle B \cong \angle C$.

Theorem 50, which concerns the "Side Angle Side" condition for triangle congruence can be used to establish several theorems pertaining to angle congruence. These theorems are similar to some axioms and theorems which were stated in the preceding section on segment congruence.

Theorem 52: If $\angle ABC \cong \angle DEF$ and halfline $\overrightarrow{BG}$ is between halflines $\overrightarrow{BA}$ and $\overrightarrow{BC}$, then there exists a unique halfline $\overrightarrow{EH}$ such that $\overrightarrow{EH}$ is between $\overrightarrow{ED}$ and $\overrightarrow{EF}$, $\angle ABG \cong \angle DEH$, and $\angle GBC \cong \angle HEF$.

Proof: Suppose $\angle ABC \cong \angle DEF$ and halfline $\overrightarrow{BG}$ is between halflines $\overrightarrow{BA}$ and $\overrightarrow{BC}$.

By Axiom 12, there exist points D^* and F^* on $\overrightarrow{ED}$ and $\overrightarrow{EF}$, respectively, such that $(B,A) \cong (E,D^*)$ and $(B,C) \cong (E,F^*)$. (See **Figure 6.7**.)

From Definition 12, A, B, C are distinct and noncollinear, as are D^*, E, F^*. Thus, by Definition 8, there are triangles $\triangle ABC$ and $\triangle D^*EF^*$.

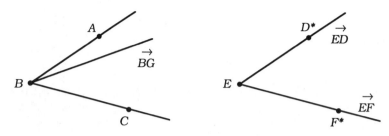

Figure 6.7

Hence, $\triangle ABC \cong \triangle D^*EF^*$, by Theorem 50, since $(B,A) \cong (E,D^*)$, $\angle ABC \cong \angle D^*EF^*$, and $(B,C) \cong (E,F^*)$. By Definition 16, $(A,C) \cong (D^*,F^*)$, $\angle BAC \cong \angle ED^*F^*$, and $\angle BCA \cong \angle EF^*D^*$. (See **Figure 6.8**.)

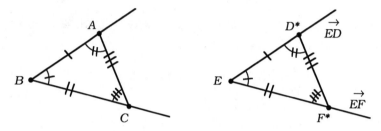

Figure 6.8

Since $\overrightarrow{BG}$ is between $\overrightarrow{BA}$ and $\overrightarrow{BC}$, then $\overrightarrow{BG} \subseteq int(\angle ABC)$, by Definition 15. Thus, $G \in int(\angle ABC)$. By Theorem 41, $\overrightarrow{BG}$ intersects (A,C) at a point K; $K \in int(\angle ABC)$. (See **Figure 6.9**.)

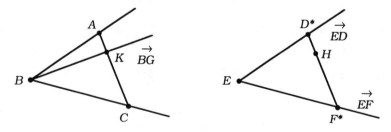

Figure 6.9

Since $A \bullet K \bullet C$ and $(A,C) \cong (D^*,F^*)$, then, by Theorem 48, there exists a unique point H on (D^*,F^*) such that $(A,K) \cong (D^*,H)$ and $D^* \bullet H \bullet F^*$. By

Theorem 38, $H \in int(\angle D^* EF^*)$.

Since $K \in int(\angle ABC)$ and $H \in int(\angle D^* EF^*)$, then, by Definition 13, A, B, K are distinct and noncollinear, as are D^*, E, H. Thus, by Definition 17, there are triangles $\triangle ABK$ and $\triangle D^*EH$.

Hence, $\triangle ABK \cong \triangle D^*EH$, by Theorem 50, since $(B,A) \cong (E,D^*)$, $\angle BAK \cong \angle ED^*H$, $(A,K) \cong (D^*,H)$. By Definition 16, $\angle ABK \cong \angle D^*EH$. Hence, $\angle ABG \cong \angle DEH$. (See **Figure 6.10**.)

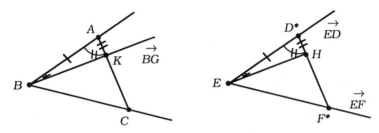

Figure 6.10

Analogously, using Theorem 47 and $\triangle BKC$ and $\triangle EHF^*$, it can be shown that $\angle GBC \cong \angle HEF$.

Finally, by Axiom 15, $\vec{EH}$ is unique. ∎

Theorem 53: If $\angle ABC$, $\angle DEF$, $\angle ABG$, $\angle DEH$, $\angle GBC$, $\angle HEF$ are angles such that halfline $\vec{BG}$ is between halflines $\vec{BA}$ and $\vec{BC}$, halfline $\vec{EH}$ is between halflines $\vec{ED}$ and $\vec{EF}$, $\angle ABG \cong \angle DEH$, and $\angle GBC \cong \angle HEF$, then $\angle ABC \cong \angle DEF$.

Theorem 54: If $\angle ABC$, $\angle DEF$, $\angle ABG$, $\angle DEH$, $\angle GBC$, $\angle HEF$ are angles such that halfline $\vec{BG}$ is between halflines $\vec{BA}$ and $\vec{BC}$, halfline $\vec{EH}$ is between halflines $\vec{ED}$ and $\vec{EF}$, $\angle ABC \cong \angle DEF$, and $\angle ABG \cong \angle DEH$, then $\angle GBC \cong \angle HEF$.

Like Theorem 50, the next two theorems are concerned with conditions which cause one triangle to be congruent to another triangle. They can also be restated in a manner similar to the restatement of Theorem 50.

Theorem 55: If $\triangle ABC$ and $\triangle DEF$ are triangles such that $\angle A \cong \angle D$,

$(A,C) \cong (D,F)$, and $\angle C \cong \angle F$, then $\triangle ABC \cong \triangle DEF$.

Language Rule 15: The preceding theorem can be called the **Angle Side Angle** congruence theorem and restated as "If two angles and the included side of one triangle are congruent, respectively, to two angles and the included side of another triangle, then the triangles are congruent."

Theorem 56: If $\triangle ABC$ and $\triangle DEF$ are two triangles such that $(A,B) \cong (D,E)$, $(B,C) \cong (E,F)$, and $(A,C) \cong (D,F)$, then $\triangle ABC \cong \triangle DEF$.

Proof: Suppose $\triangle ABC$ and $\triangle DEF$ are two triangles such that $(A,B) \cong (D,E)$, $(B,C) \cong (E,F)$, and $(A,C) \cong (D,F)$.

By Definition 8, B is not on line AC.

By Axiom 15, there exists a halfline $\overrightarrow{AG}$, on the side of line AC which does not contain B, such that $\angle CAG \cong \angle D$. By Axiom 12, there exists a point H on $\overrightarrow{AG}$ such that $(A,H) \cong (D,E)$.

By Theorem 21 and Theorem 22, H and B are on opposite sides of AC, and A, C, H are distinct and noncollinear. Thus, A, C, H are the vertices of $\triangle ACH$.

Since $(A,H) \cong (D,E)$, $\angle CAH \cong \angle D$, and $(A,C) \cong (D,F)$, then, by Theorem 50, $\triangle ACH \cong \triangle DFE$. By Definition 16, $(H,C) \cong (E,F)$. By Axiom 13, $(A,B) \cong (A,H)$ and $(B,C) \cong (H,C)$, since $(A,B) \cong (D,E)$ and $(B,C) \cong (E,F)$.

By Theorem 20 and Definition 4, there exists a point J on AC such that $B \bullet J \bullet H$. By Definition 5 and Theorem 15, J is a point on line BH, which is distinct from AC, since B is not on AC.

Since A, C, J are all on line AC, then either $J \in \{A,C\}$ or $J \notin \{A,C\}$. If $J \in \{A,C\}$, then either $J = A$ or $J = C$. If $J \notin \{A,C\}$, then J, A, C are distinct and collinear, and, by Axiom 8, exactly one of $A \bullet J \bullet C$, $J \bullet A \bullet C$, $A \bullet C \bullet J$ occurs. Hence, there are five possibilities to be considered. (See **Figure 6.11**.)

Claim: If $A \bullet J \bullet C$, then $\triangle ABC \cong \triangle DEF$.

Proof: Suppose $A \bullet J \bullet C$.

Since A, $C \ne J$ and $BH \ne AC$, then, by Theorem 2, A, $C \notin BH$. Hence, A, B, H are distinct and noncollinear, as are C, B, H. Thus, A, B, H are the vertices of $\triangle ABH$, and C, B, H are the vertices of $\triangle CBH$.

In $\triangle ABH$, $(A,B) \cong (A,H)$, and in $\triangle CBH$, $(C,B) \cong (C,H)$. Thus, by Theo-

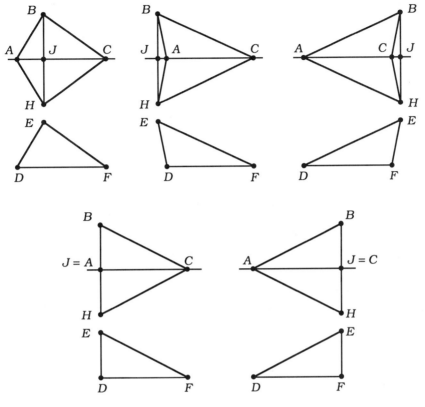

Figure 6.11

rem 51, ∠ABH ≅ ∠AHB and ∠CBH ≅ ∠CHB.

Since A and C are points on two sides of ∠ABC, and J is a point such that A•J•C, then, by Theorem 38, J ∈ int(∠ABC). By Theorem 40, halfline $\vec{BJ}$ is a subset of int(∠ABC), and, by Definition 15, $\vec{BJ}$ is between $\vec{BA}$ and $\vec{BC}$. Similarly, $\vec{HJ}$ is between $\vec{HA}$ and $\vec{HC}$.

Thus, by Theorem 53, ∠ABC ≅ ∠AHC.

Since (A,B) ≅ (A,H), ∠ABC ≅ ∠AHC, and (B,C) ≅ (H,C), then, by Theorem 50, △ABC ≅ △AHC.

Therefore, by Theorem 49, △ABC ≅ △DEF, since △ABC ≅ △AHC and △AHC ≅ △DEF. ♦

Claim: If J•A•C, then △ABC ≅ △DEF.

Proof: Left as an exercise. ♦

Claim: If $A \bullet C \bullet J$, then $\triangle ABC \cong \triangle DEF$.

Proof: Analogous to the preceding **Claim**. ♦

Claim: If $J = A$, then $\triangle ABC \cong \triangle DEF$.

Proof: Left as an exercise. ♦

Claim: If $J = C$, then $\triangle ABC \cong \triangle DEF$.

Proof: Analogous to the preceding **Claim**. ♦

Thus, if $\triangle ABC$ and $\triangle DEF$ are two triangles such that $(A,B) \cong (D,E)$, $(B,C) \cong (E,F)$, and $(A,C) \cong (D,F)$, then $\triangle ABC \cong \triangle DEF$. ∎

Language Rule 16: The preceding theorem can be called the **Side Side Side** congruence theorem and restated as "If the sides of one triangle are congruent, respectively, to the sides of another triangle, then the triangles are congruent."

Exercises 6.3

1. Develop a proof for Theorem 49.

2. Develop a proof for Theorem 51.

3. Develop a proof for Theorem 53.

4. Develop a proof for Theorem 54.

5. Develop a proof for Theorem 55.

6. Complete the proof for Theorem 56.

7. Develop a proof for this statement:
   If $\triangle ABC$ is a triangle such that $\angle B \cong \angle C$, then $(A,B) \cong (A,C)$.

6.4 Comparison of Segments

The two concepts of betweenness for points and congruence for segments can be combined to develop a definition which can be used for comparing segments. This definition can be used, along with some preceding theorems, to obtain several theorems pertaining to the comparison of segments.

Definition 18: Segment (A,B) is **less than** segment (C,D) if there exists a point E such that $C \bullet E \bullet D$ and $(A,B) \cong (C,E)$.

Notation Rule 15: (A,B) is less than (C,D) will be symbolized as $(A,B) < (C,D)$.

Theorem 57: If (A,B), (C,D), (E,F) are segments such that $(A,B) < (C,D)$ and $(A,B) \cong (E,F)$, then $(E,F) < (C,D)$.

Proof: Suppose (A,B), (C,D), (E,F) are segments such that $(A,B) < (C,D)$ and $(A,B) \cong (E,F)$.
Since $(A,B) < (C,D)$, then, by Definition 18, there exists a point G such that $C \bullet G \bullet D$ and $(A,B) \cong (C,G)$.
Since $(A,B) \cong (E,F)$ and $(A,B) \cong (C,G)$, then, by Axiom 13, $(E,F) \cong (C,G)$.
Thus, G is a point such that $C \bullet G \bullet D$ and $(E,F) \cong (C,G)$.
By Definition 18, $(E,F) < (C,D)$. ∎

Theorem 58: If (A,B), (C,D), (E,F) are segments such that $(A,B) < (C,D)$ and $(C,D) \cong (E,F)$, then $(A,B) < (E,F)$.

Theorem 59: If (A,B), (C,D), (E,F) are segments such that $(A,B) < (C,D)$ and $(C,D) < (E,F)$, then $(A,B) < (E,F)$.

Theorem 60: If (A,B) and (C,D) are two segments, then exactly one of the following is true:
(a) $(A,B) < (C,D)$;
(b) $(A,B) \cong (C,D)$;
(c) $(C,D) < (A,B)$.

Proof: Suppose (A,B) and (C,D) are two segments.

By Theorem 15, (C,D) is on line CD. By Axiom 12, there exists on $\overrightarrow{CD}$ exactly one point E such that $(A,B) \equiv (C,E)$.

Since E is on $\overrightarrow{CD}$, then E is on line CD and either $E = D$ or $E \neq D$. If $E \neq D$, then, by Axiom 8, exactly one of $E \bullet C \bullet D$, $C \bullet E \bullet D$, $C \bullet D \bullet E$ occurs. $E \bullet C \bullet D$ is not possible, by Definition 2, since E is on the D-side of C.

Thus, there are exactly three possibilities: $E = D$, $C \bullet E \bullet D$, $C \bullet D \bullet E$.

Claim: $E = D$ if and only if $(A,B) \equiv (C,D)$.

Subclaim: If $E = D$, then $(A,B) \equiv (C,D)$.

Proof: Suppose $E = D$.
Since $(A,B) \equiv (C,E)$ and $E = D$, then $(A,B) \equiv (C,D)$. ♦♦

Subclaim: If $(A,B) \equiv (C,D)$, then $E = D$.

Proof: Suppose $(A,B) \equiv (C,D)$.
Since $(A,B) \equiv (C,D)$ and $(A,B) \equiv (C,E)$, by Axiom 13, $(C,D) \equiv (C,E)$.
Since E is on $\overrightarrow{CD}$, then, by Axiom 12, $E = D$. ♦♦

Thus, $E = D$ if and only if $(A,B) \equiv (C,D)$. ♦

Claim: $C \bullet E \bullet D$ if and only if $(A,B) < (C,D)$.

Subclaim: If $C \bullet E \bullet D$, then $(A,B) < (C,D)$.

Proof: Suppose $C \bullet E \bullet D$.
Since $(A,B) \equiv (C,E)$, then, by Definition 18, $(A,B) < (C,D)$. ♦♦

Subclaim: If $(A,B) < (C,D)$, then $C \bullet E \bullet D$.

Proof: Suppose $(A,B) < (C,D)$.
Since $(A,B) < (C,D)$, then, by Definition 18, there exists a point F such that $C \bullet F \bullet D$ and $(A,B) \equiv (C,F)$.

Since $C \bullet F \bullet D$, then, by Definition 2, F is on $\overrightarrow{CD}$.
By Axiom 13, $(C,E) \cong (C,F)$, since $(A,B) \cong (C,E)$ and $(A,B) \cong (C,F)$.

Hence, $E = F$, by Axiom 12, since each of E and F is on $\overrightarrow{CD}$ and $(C,E) \cong (C,F)$.

Thus, $C \bullet F \bullet D$ and $F = E$. Hence, $C \bullet E \bullet D$. ♦♦

Therefore, $C \bullet E \bullet D$ if and only if $(A,B) < (C,D)$. ♦

Claim: $C \bullet D \bullet E$ if and only if $(C,D) < (A,B)$.

Subclaim: If $C \bullet D \bullet E$, then $(C,D) < (A,B)$.

Proof: Suppose $C \bullet D \bullet E$.
Since $(C,E) \cong (A,B)$, then, by Theorem 48, there exists a point F such that $A \bullet F \bullet B$ and $(C,D) \cong (A,F)$.

Hence, $(C,D) \cong (A,F)$ and $A \bullet F \bullet B$. Therefore, by Definition 18, $(C,D) < (A,B)$. ♦♦

Subclaim: If $(C,D) < (A,B)$, then $C \bullet D \bullet E$.

Proof: Suppose $(C,D) < (A,B)$.
By Definition 18, there exists a point F such that $A \bullet F \bullet B$ and $(C,D) \cong (A,F)$.

By Axiom 12, there exists a point G, on line CD with $C \bullet D \bullet G$, such that $(D,G) \cong (F,B)$.

Hence, $A \bullet F \bullet B$, $C \bullet D \bullet G$, $(A,F) \cong (C,D)$, and $(D,G) \cong (F,B)$. Thus, by Axiom 14, $(A,B) \cong (C,G)$.

Since $(A,B) \cong (C,E)$ and $(A,B) \cong (C,G)$, then, by Axiom 13, $(C,E) \cong (C,G)$.

Thus, $(C,E) \cong (C,G)$ and each of E and G is on $\overrightarrow{CD}$. By Axiom 12, $E = G$. Hence, $C \bullet D \bullet G$ and $E = G$. Therefore, $C \bullet D \bullet E$. ♦♦

Thus, $C \bullet D \bullet E$ if and only $(C,D) < (A,B)$. ♦

Therefore, exactly one of $(A,B) < (C,D)$, $(A,B) \cong (C,D)$, $(C,D) < (A,B)$ will occur. ■

Exercises 6.4

1. Develop a proof for Theorem 58.

2. Develop a proof for Theorem 59.

6.5 Comparison of Angles

Angles can be compared in much the same manner as that used for comparing segments. Since their proofs are somewhat similar to those of the theorems in the preceding section, the theorems pertaining to angle comparison will all be stated without proof.

Definition 19: Angle $\angle ABC$ is **less than** angle $\angle DEF$ if there exists a halfline $\overrightarrow{EG}$ such that $\overrightarrow{EG}$ is between $\overrightarrow{ED}$ and $\overrightarrow{EF}$ and $\angle ABC \cong \angle GEF$.

Notation Rule 16: $\angle ABC$ is **less than** $\angle DEF$ will be symbolized as $\angle ABC < \angle DEF$.

Theorem 61: If $\angle ABC$, $\angle DEF$, $\angle GHI$ are angles such that $\angle ABC < \angle DEF$ and $\angle ABC \cong \angle GHI$, then $\angle GHI < \angle DEF$.

Theorem 62: If $\angle ABC$, $\angle DEF$, $\angle GHI$ are angles such that $\angle ABC < \angle DEF$ and $\angle DEF \cong \angle GHI$, then $\angle ABC < \angle GHI$.

Theorem 63: If $\angle ABC$, $\angle DEF$, $\angle GHI$ are angles such that $\angle ABC < \angle DEF$ and $\angle DEF < \angle GHI$, then $\angle ABC < \angle GHI$.

Theorem 64: If $\angle ABC$ and $\angle DEF$ are two angles, then exactly one of the following is true:
(a) $\angle ABC < \angle DEF$;
(b) $\angle DEF < \angle ABC$;
(c) $\angle ABC \cong \angle DEF$.

Exercises 6.5

1. Develop a proof for Theorem 61.

2. Develop a proof for Theorem 62.

3. Develop a proof for Theorem 63.

4. Develop a proof for Theorem 64.

6.6 Some Additional Theorems Concerning Angle Congruences and Comparisons

This section contains a collection of theorems concerning congruences and comparisons involving some special types of angles.

Definition 20: Two angles having the same vertex and one side in common while the sides not in common are on opposite sides of the common side are called **adjacent angles**. (See **Figure 6.12**.)

Figure 6.12

Definition 21: Adjacent angles whose sides not in common are opposite halflines are called a **linear pair of angles**. (See **Figure 6.13**.)

Figure 6.13

Theorem 65: If two angles are congruent, then the angles with which they form linear pairs are congruent.

Proof: Suppose ∠ABC and ∠CBG are a linear pair of angles, ∠DEF and ∠FEH are a linear pair of angles, and ∠ABC ≅ ∠DEF. (See **Figure 6.14**.)

By Axiom 12, there exists points J, K, L on $\overrightarrow{EH}$, $\overrightarrow{EF}$, $\overrightarrow{ED}$, respectively, such that (E,J) ≅ (B,G), (E,K) ≅ (B,C), (E,L) ≅ (B,A). (See **Figure 6.15**.)

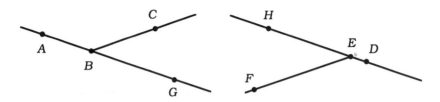

Figure 6.14

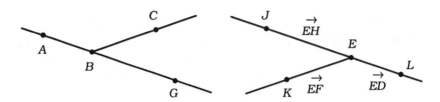

Figure 6.15

By Definition 12, $\overrightarrow{BC}$ is not collinear with $\overrightarrow{BA}$ or $\overrightarrow{BG}$. Hence, C is not on line AG, by Theorem 2, since B is on AG. Thus, A, B, C are distinct and noncollinear, and, by Definition 8, are the vertices of $\triangle ABC$. Similarly, B, G, C are the vertices of $\triangle BGC$, E, J, K are the vertices of $\triangle EJK$, and L, E, K are the vertices of $\triangle KEL$. (See **Figure 6.16**.)

Figure 6.16

Since, $(A,B) \cong (L,E)$, $\angle ABC \cong \angle LEK$, and $(B,C) \cong (E,K)$, then, by Theorem 49, $\triangle ABC \cong \triangle LEK$. Hence, by Definition 16, $(A,C) \cong (L,K)$ and $\angle BAC \cong \angle ELK$.

Since $(A,B) \cong (L,E)$ and $(B,G) \cong (E,J)$, then, by Axiom 14, $(A,G) \cong (L,J)$.

Thus, $\triangle ACG \cong \triangle LKJ$, by Theorem 50. Hence, by Definition 16, $(C,G) \cong (K,J)$ and $\angle BGC \cong \angle EJK$.

By Theorem 50, $\triangle BCG \cong \triangle EKJ$.

Hence, $\angle CBG \cong \angle KEJ$, by Definition 16. ∎

Definition 22: Two angles which are congruent, respectively, to a linear pair of angles are called **supplementary angles**.

Language Rule 17: Each of a pair of supplementary angles is called the **supplement** of the other.

Theorem 66: If two angles are congruent, then their supplementary angles are congruent.

Theorem 67: If two angles are adjacent and supplementary, then they are a linear pair of angles.

Proof: Suppose ∠AOB and ∠AOC are adjacent, with $\overrightarrow{OA}$ being the common side and $\overrightarrow{OB}$, $\overrightarrow{OC}$ being on opposite sides of $\overrightarrow{OA}$; suppose further that ∠AOB and ∠AOC are supplementary and $\overrightarrow{OD}$ is the halfline opposite $\overrightarrow{OB}$. (See **Figure 6.17**.)

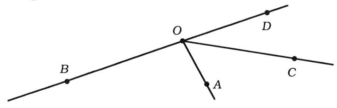

Figure 6.17

Since ∠AOB and ∠AOD are adjacent angles whose sides not in common are opposite halflines, they are a linear pair of angles, by Definition 21. By Axiom 16, ∠AOB ≅ ∠AOB and ∠AOD ≅ ∠AOD. Thus, by Definition 22, ∠AOB and ∠AOD are supplementary.

Hence, ∠AOB and ∠AOC are supplementary, ∠AOB and ∠AOD are supplementary, and ∠AOB ≅ ∠AOB. Thus, by Theorem 66, ∠AOC ≅ ∠AOD.

Since $\overrightarrow{OB}$, $\overrightarrow{OC}$ are on opposite sides of $\overrightarrow{OA}$ and $\overrightarrow{OB}$, $\overrightarrow{OD}$ are on opposite sides of $\overrightarrow{OA}$, then, by Theorem 21, B, C are on opposite sides of $\overrightarrow{OA}$ and B, D are on opposite sides of $\overrightarrow{OA}$. By Theorem 22, C, D are on the same side of $\overrightarrow{OA}$, and, by Theorem 21, $\overrightarrow{OC}$, $\overrightarrow{OD}$ are on the same side of $\overrightarrow{OA}$.

By Axiom 15, $\overrightarrow{OC} = \overrightarrow{OD}$, since $\angle AOC \cong \angle AOD$ and $\overrightarrow{OC}$, $\overrightarrow{OD}$ are on the same side of $\overrightarrow{OA}$. Hence, $\overrightarrow{OB}$ and $\overrightarrow{OC}$ are opposite halflines.

Thus, $\angle AOB$ and $\angle AOC$ are adjacent angles whose sides not in common are opposite halflines.

Hence, by Definition 21, $\angle AOB$ and $\angle AOC$ are a linear pair of angles. ∎

Definition 23: Two angles having the same vertex and whose sides are two pairs of opposite halflines are called **vertical angles**. (See **Figure 6.18**.)

Figure 6.18

Theorem 68: Vertical angles are congruent.

Definition 24: An angle which is congruent to its supplementary angle is called a **right angle**.

Theorem 69: An angle congruent to a right angle is a right angle.

Theorem 70: If two angles are right angles, then they are congruent.

Proof: Suppose $\angle ABC$ and $\angle EFG$ are two right angles.

Points A and B lie on a line m and points E and F lie on a line n, by Axiom 1. By Axiom 5, Axiom 6, and Axiom 1b, there exist points D and H on lines m and n, respectively, such that $A \bullet B \bullet D$ and $E \bullet F \bullet H$. (See **Figure 6.19**.)

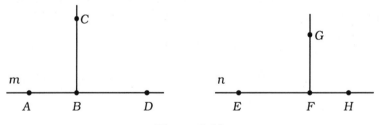

Figure 6.19

By Definition 2, $\overrightarrow{BA}$, $\overrightarrow{BD}$ are opposite halflines and $\overrightarrow{FE}$, $\overrightarrow{FH}$ are opposite halflines.

By Definition 21, Axiom 16, and Definition 22, $\angle ABC$, $\angle DBC$ are supplementary angles and $\angle EFG$, $\angle HFG$ are supplementary angles. By Definition 24, $\angle ABC \cong \angle DBC$ and $\angle EFG \cong \angle HFG$.

By Theorem 64, exactly one of $\angle ABC < \angle EFG$, $\angle EFG < \angle ABC$, and $\angle ABC \cong \angle EFG$ is true.

Claim: $\angle ABC < \angle EFG$ is not true.

Proof: Suppose $\angle ABC < \angle EFG$.

By Definition 19, there exists a halfline $\overrightarrow{FJ}$ such that $\overrightarrow{FJ}$ is between $\overrightarrow{FE}$ and $\overrightarrow{FG}$ and $\angle ABC \cong \angle EFJ$. (See **Figure 6.20**.)

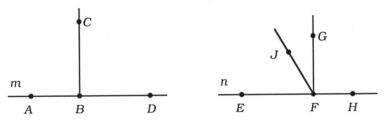

Figure 6.20

Angles $\angle EFJ$ and $\angle HFJ$ are supplementary angles, by Definition 21, Axiom 16, and Definition 22. Since $\angle ABC \cong \angle EFJ$, then $\angle DBC \cong \angle HFJ$, by Theorem 66.

Thus, $\angle ABC \cong \angle DBC$, $\angle ABC \cong \angle EFJ$, and $\angle DBC \cong \angle HFJ$. By Axiom 16, $\angle EFJ \cong \angle HFJ$.

Since $\angle EFG \cong \angle HFG$ and $\overrightarrow{FJ}$ is between $\overrightarrow{FE}$ and $\overrightarrow{FG}$, then, by Theorem 52, there exists a unique halfline $\overrightarrow{FK}$ such that $\overrightarrow{FK}$ is between $\overrightarrow{FG}$ and $\overrightarrow{FH}$ and $\angle EFJ \cong \angle HFK$. (See **Figure 6.21**.)

Thus, $\angle EFJ \cong \angle HFJ$ and $\angle EFJ \cong \angle HFK$. By Axiom 16, $\angle HFJ \cong \angle HFK$.

Since $\overrightarrow{FE}$ and $\overrightarrow{FH}$ are opposite halflines and $\overrightarrow{FJ}$ is between $\overrightarrow{FE}$ and $\overrightarrow{FG}$, then, by Theorem 44, $\overrightarrow{FG}$ is between $\overrightarrow{FJ}$ and $\overrightarrow{FH}$. Thus, by Definition 19, $\angle HFG < \angle HFJ$.

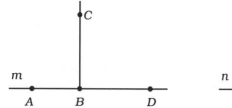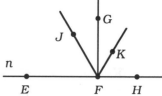

Figure 6.21

Since $\vec{FK}$ is between $\vec{FG}$ and $\vec{FH}$, then, by Definition 19, ∠HFK < ∠HFG.
Thus, ∠HFG < ∠HFJ and ∠HFK < ∠HFG. Hence, by Theorem 63, ∠HFK < ∠HFJ.

Since ∠HFJ ≅ ∠HFK and ∠HFK < ∠HFJ, then Theorem 64 is contradicted.

Hence, ∠ABC < ∠EFG is not true. ♦

Claim: ∠EFG < ∠ABC is not true.

Proof: Analogous to the preceding **Claim.** ♦

Therefore, each of ∠ABC < ∠EFG and ∠EFG < ∠ABC is not true. Hence, by Theorem 64, ∠ABC ≅ ∠EFG. ∎

Definition 25: An angle which is less than a right angle is called an **acute angle**.

Definition 26: An angle which is neither a right angle nor an acute angle is called an **obtuse angle**.

Theorem 71: If one of two angles is a right angle and the other is an obtuse angle, then the right angle is less than the obtuse angle.

Theorem 72: If one of two angles is an acute angle and the other is an obtuse angle, then the acute angle is less than the obtuse angle.

Proof: Suppose ∠ABC and ∠DEF are two angles such that ∠ABC is an acute angle and ∠DEF is an obtuse angle; suppose further that ∠LMN and ∠NMP are congruent and form a linear pair of angles.

By Axiom 16 and Definition 22, ∠LMN and ∠NMP are supplementary angles. Hence, by Definition 24, each of ∠LMN and ∠NMP is a right angle.

Since ∠ABC is an acute angle, then, by Definition 25, ∠ABC < ∠LMN. Since ∠DEF is an obtuse angle, then ∠LMN < ∠DEF, by Theorem 71. Thus, ∠ABC < ∠LMN and ∠LMN < ∠DEF. By Theorem 63, ∠ABC < ∠DEF. ∎

Theorem 73: For a pair of supplementary angles, one is an acute angle if and only if the other is an obtuse angle.

Proof: Suppose ∠ABC and ∠DEF are a pair of supplementary angles; suppose further that ∠LMN and ∠NMP are congruent and form a linear pair of angles.

By Axiom 16 and Definition 22, ∠LMN and ∠NMP are supplementary angles. Hence, by Definition 24, each of ∠LMN and ∠NMP is a right angle. By Definition 21, $\overrightarrow{ML}$ and $\overrightarrow{MP}$ are opposite halflines.

Claim: If one of ∠ABC and ∠DEF is an acute angle, then the other is an obtuse angle.

Proof: Suppose one of ∠ABC and ∠DEF is an acute angle; without loss of generality, suppose ∠ABC is an acute angle.

By Definition 25, ∠ABC < ∠LMN. Hence, by Definition 19, there exists a halfline $\overrightarrow{MQ}$ such that $\overrightarrow{MQ}$ is between $\overrightarrow{ML}$ and $\overrightarrow{MN}$ and ∠ABC ≅ ∠LMQ. (See **Figure 6.22**.)

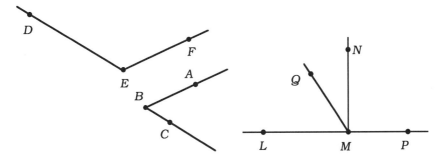

Figure 6.22

By Definition 21, Axiom 16, and Definition 22, ∠LMQ and ∠QMP are supplementary angles.

Since ∠ABC and ∠DEF are supplementary, ∠LMQ and ∠QMP are supplementary, and ∠ABC ≅ ∠LMQ, then, by Theorem 66, ∠DEF ≅ ∠QMP.

Since $\overrightarrow{ML}$ and $\overrightarrow{MP}$ are opposite halflines and $\overrightarrow{MQ}$ is between $\overrightarrow{ML}$ and $\overrightarrow{MN}$, then, by Theorem 44, $\overrightarrow{MN}$ is between $\overrightarrow{MQ}$ and $\overrightarrow{MP}$. Hence, $\angle NMP < \angle QMP$, by Definition 19.

Thus, by Theorem 62, $\angle NMP < \angle DEF$, since $\angle NMP < \angle QMP$ and $\angle DEF \cong \angle QMP$.

Hence, by Theorem 64, $\angle NMP \not\cong \angle DEF$ and $\angle DEF \not< \angle NMP$. Thus, by Theorem 70, $\angle DEF$ is not a right angle, and, by Definition 25, $\angle DEF$ is not an acute angle.

Therefore, by Definition 26, $\angle DEF$ is obtuse. ♦

Claim: If one of $\angle ABC$ and $\angle DEF$ is an obtuse angle, then the other is an acute angle.

Proof: Left as an exercise. ♦

Therefore, for a pair of supplementary angles, one is an acute angle if and only if the other is an obtuse angle. ∎

Theorem 74: If two intersecting lines contain one right angle, then they contain four right angles.

Theorem 75: If two angles are supplementary and not congruent, then one of them is an acute angle.

Proof: Suppose $\angle ABC$ and $\angle DEF$ are supplementary and not congruent.

By Theorem 64, $\angle ABC < \angle DEF$ or $\angle DEF < \angle ABC$.

Claim: If $\angle ABC < \angle DEF$, then $\angle ABC$ is acute.

Proof: Suppose $\angle ABC$ is not acute; suppose further that $\angle LMN$ and $\angle NMP$ are congruent and form a linear pair of angles.

By Axiom 16 and Definition 22, $\angle LMN$ and $\angle NMP$ are supplementary angles. Hence, by Definition 24, each of $\angle LMN$ and $\angle NMP$ is a right angle.

Since $\angle ABC$ is not an acute angle, then, by Definition 25, $\angle ABC \not< \angle LMN$. Hence, by Theorem 64, $\angle ABC \cong \angle LMN$ or $\angle LMN < \angle ABC$.

Subclaim: If $\angle ABC \cong \angle LMN$, then $\angle ABC \not< \angle DEF$.

Proof: Left as an exercise. ♦♦

Subclaim: If ∠LMN < ∠ABC, then ∠ABC ≮ ∠DEF.

Proof: Left as an exercise. ♦♦

Therefore, if ∠ABC < DEF, then ∠ABC is acute. ♦

Claim: If ∠DEF < ∠ABC, then ∠DEF is acute.

Proof: Analogous to the preceding **Claim**. ♦

Therefore, if two angles are supplementary and not congruent, then one of them is an acute angle. ∎

Exercises 6.6

1. Develop a proof for Theorem 66.
2. Develop a proof for Theorem 68.
3. Develop a proof for Theorem 69.
4. Develop a proof for Theorem 71.
5. Complete the proof for Theorem 73.
6. Develop a proof for Theorem 74.
7. Complete the proof for Theorem 75.

****

SUPPLEMENTARY TOPICS

CONSTRUCTIONS

24. Construct a triangle $\triangle ABC$ such that $\angle A$ is congruent to the given $\angle D$, the altitude to side $\overline{AB}$ is congruent to the given segment $\overline{EF}$, and the altitude to side $\overline{AC}$ is congruent to the given segment $\overline{GH}$.

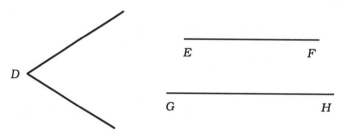

25. Construct $\angle ABC$ so that $\angle ABC \cong \angle PQR$, which is given, and $\overline{AB} \cong \overline{PQ}$. Construct a circle which will be tangent to line BC at point B and pass through point A.

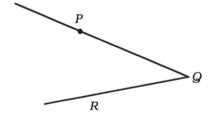

26. Construct a triangle $\triangle ABC$ such that side $\overline{AB}$ is congruent to the given segment $\overline{DE}$, side $\overline{BC}$ is congruent to the given segment $\overline{FG}$, and the median to side $\overline{AC}$ is congruent to the segment $\overline{HJ}$.

27. Using an initial segment measuring 1 inch, select any three noncollinear points A, B, C, such that they are located between 2 and 3 inches apart. Determine all points which are equidistant from A and B and also 1.5 inches from C.

28. Construct two parallel lines which are located about 2 inches apart, and select a point *A* which is located anywhere between the lines. Determine a circle which is tangent to both lines and passes through *A*. How many such circles are possible?

*****

PROBLEMS

13. The bisectors of two exterior angles of a triangle intersect to form an angle. Show that this angle has measure equal to one-half the measure of the triangle's third exterior angle.

14. In the following diagram, four congruent circles are positioned so that their centers are the vertices of a square whose sides measure 5 units. Determine the area and perimeter of the portion of the square which is common to all four circles.

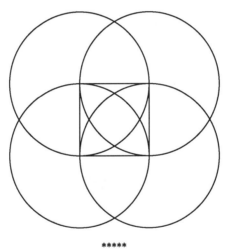

*****

TRANSFORMATION GEOMETRY

T-6: Isometry Relationships

Our study of the special collection of transformations known as

SUPPLEMENTARY TOPICS   149

isometries would be somewhat incomplete without a concluding section which presents some special characteristics and describes various relationships existing among **line reflections, rotations, translations,** and **glide reflections**.

Definition T-12: A transformation **T fixes** a point P, if $\mathbf{T}(P) = P$. The point P is called a **fixed point** for **T**.

Theorem T-9: An isometry which fixes exactly one point is a rotation.

Theorem T-10: An isometry which fixes at least two points of a line fixes every point of the line.

Theorem T-11: An isometry which fixes at least two points is a reflection or the identity transformation.

Theorem T-12: An isometry which fixes at least three noncollinear points is the identity transformation.

Theorem T-13: If A, B, C are three noncollinear points and **T** is an isometry such that $\mathbf{T}(A) = A'$, $\mathbf{T}(B) = B'$, $\mathbf{T}(C) = C'$, then $\mathbf{T}(X) = X'$ can be determined for any point X.

Theorem T-14: If A, B, C are three noncollinear points and **M** and **T** are isometries such that $\mathbf{M}(A) = \mathbf{T}(A)$, $\mathbf{M}(B) = \mathbf{T}(B)$, $\mathbf{M}(C) = \mathbf{T}(C)$, then $\mathbf{M} = \mathbf{T}$.

Remark T-6: Since isometries are special types of functions, the concept of function composition can be used with isometries.

Theorem T-15: An isometry which fixes exactly one point is equivalent to the composition of two line reflections.

Theorem T-16: An isometry which fixes at least one point is equivalent to a composition of at most two line reflections.

Theorem T-17: A composition of any finite number of reflections is an isometry.

Theorem T-18: Every isometry is equivalent to the composition of at most three line reflections.

Exercises T-6

1. How many nonfixed point-and-image pairs are necessary to determine each?
   (a) a line reflection
   (b) a rotation
   (c) a translation
   (d) a glide reflection

2. How many points does each fix?
   (a) a line reflection
   (b) a rotation
   (c) a translation
   (d) a glide reflection

3. Using a straightedge and a compass, draw a diagram similar to the one in **Figure T-53**, in which $\triangle A'B'C'$ is the image of $\triangle ABC$ under an isometry **M**. Using only a compass and straightedge, determine the image of P and the preimage of Q.

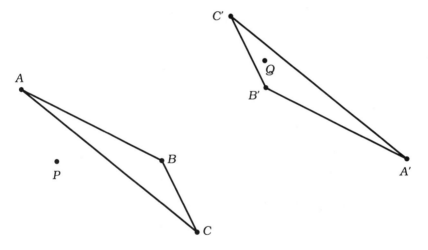

Figure T-53

4. Using a straightedge, draw a diagram similar to the one in **Figure T-54**. Use constructions to determine $\mathbf{R}_n(\mathbf{R}_m(P))$ and $\mathbf{R}_n(\mathbf{R}_m(Q))$. Use constructions to determine *one* isometry **M** such that $\mathbf{M}(X) = \mathbf{R}_n(\mathbf{R}_m(X))$ for any point X.

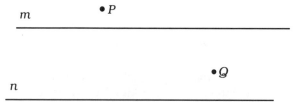

Figure T-54

5. Using a straightedge, draw a diagram similar to the one in **Figure T-55**. Use constructions to determine $\mathbf{R}_n(\mathbf{R}_m(P))$ and $\mathbf{R}_n(\mathbf{R}_m(Q))$. Use constructions to determine *one* isometry **M** such that $\mathbf{M}(X) = \mathbf{R}_n(\mathbf{R}_m(X))$ for any point X.

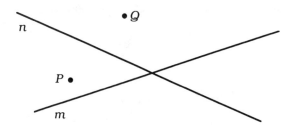

Figure T-55

6. Using a straightedge, draw a diagram similar to the one in **Figure T-56**. Use constructions to determine $\mathbf{R}_o(\mathbf{R}_n(\mathbf{R}_m(P)))$ and $\mathbf{R}_o(\mathbf{R}_n(\mathbf{R}_m(Q)))$. Use constructions to determine *one* isometry **M** such that $\mathbf{M}(X) = \mathbf{R}_o(\mathbf{R}_n(\mathbf{R}_m(X)))$ for any point X.

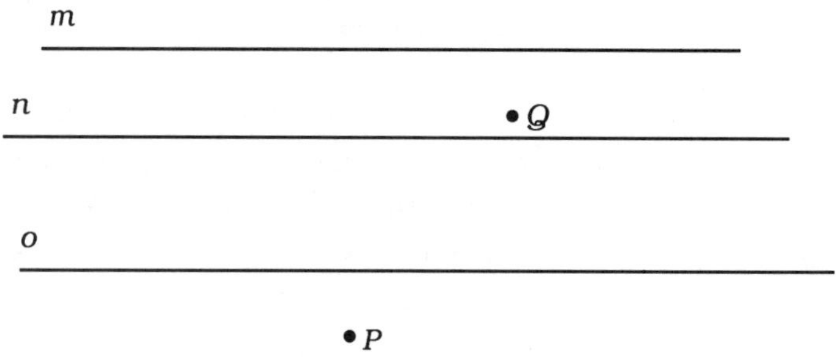

Figure T-56

7. Using a straightedge, draw a diagram similar to the one in **Figure T-57**. Use constructions to determine $R_o(R_n(R_m(P)))$ and $R_o(R_n(R_m(Q)))$. Use constructions to determine *one* isometry **M** such that $M(X) = R_o(R_n(R_m(X)))$ for any point X.

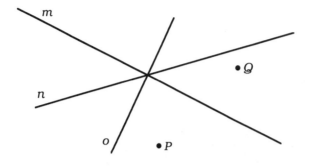

Figure T-57

8. Using a straightedge, draw a diagram similar to the one in **Figure T-58**. Use constructions to determine $R_o(R_n(R_m(P)))$ and $R_o(R_n(R_m(Q)))$. Use constructions to determine *one* isometry **M** such that $M(X) = R_o(R_n(R_m(X)))$, for any point X.

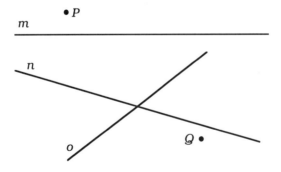

Figure T-58

9. Using a straightedge and compass, draw a diagram similar to the one in **Figure T-59**. Use constructions to determine the images of A and B and the preimage of C for the isometries composition $P_{60} \circ R_m$. Use constructions to determine *one* isometry **M** such

that $\mathbf{M}(X) = \mathbf{P}_{60}(\mathbf{R}_m(X))$, for any point X.

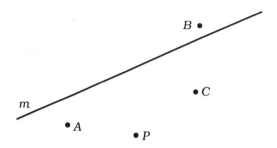

Figure T-59

10. Using a straightedge and compass, draw a diagram similar to the one in **Figure T-60**. Use constructions to determine the images of A and B and the preimage of C for the isometries composition $\mathbf{R}_n \circ \mathbf{G}$, where $\mathbf{G}$ involves line m and the given translation. Use constructions to determine *one* isometry $\mathbf{M}$ such that $\mathbf{M}(X) = \mathbf{R}_n(\mathbf{G}(X))$, for any point X.

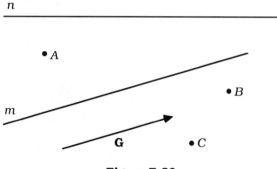

Figure T-60

11. Using a straightedge and compass, draw a diagram similar to the one in **Figure T-61**. Use constructions to determine the images of A and B and the preimage of C for the isometries composition $\mathbf{G}_n \circ \mathbf{G}_m$, where $\mathbf{G}_n$ involves line n and the given translation, while $\mathbf{G}_m$ involves line m and the given translation. Use constructions to determine *one* isometry $\mathbf{M}$ such that $\mathbf{M}(X) = \mathbf{G}_n(\mathbf{G}_m(X))$, for any point X.

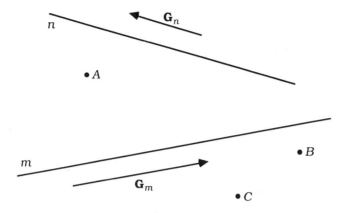

Figure T-61

12. Using a straightedge and compass, draw a diagram similar to the one in **Figure T-62**. Use constructions to determine the images of A and B and the preimage of C for the isometries composition $\mathbf{P}_{45} \circ \mathbf{Q}_{60}$. Use constructions to determine *one* isometry **M** such that $\mathbf{M}(X) = \mathbf{P}_{45}(\mathbf{Q}_{60}(X))$, for any point X.

Figure T-62

13. Using a straightedge and compass, draw a diagram similar to the one in **Figure T-63**. Use constructions to determine the location of a line m such that $\mathbf{R}_m(A) = A'$, $\mathbf{R}_m(B) = B'$, $\mathbf{R}_m(C) = C'$.

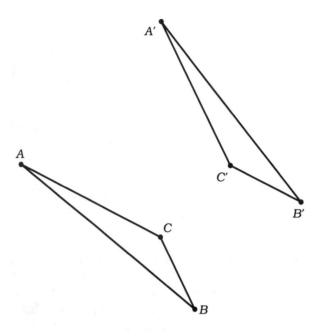

Figure T-63

14. Using a straightedge and compass, draw a diagram similar to the one in **Figure T-64**. Use constructions to determine the location of lines m and n such that $\mathbf{R}_n(\mathbf{R}_m(A)) = A'$, $\mathbf{R}_n(\mathbf{R}_m(B)) = B'$, $\mathbf{R}_n(\mathbf{R}_m(C)) = C'$.

15. Using a straightedge and compass, draw a diagram similar to the one in **Figure T-65**. Use constructions to determine the location of lines m, n, o such that $\mathbf{R}_o(\mathbf{R}_n(\mathbf{R}_m(A))) = A'$, $\mathbf{R}_o(\mathbf{R}_n(\mathbf{R}_m(B))) = B'$, $\mathbf{R}_o(\mathbf{R}_n(\mathbf{R}_m(C))) = C'$.

16. Develop an informal proof for Theorem T-9.

17. Develop an informal proof for Theorem T-10.

18. Develop an informal proof for Theorem T-11.

19. Develop an informal proof for Theorem T-12.

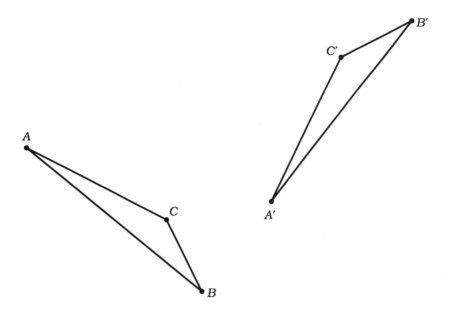

Figure T-64

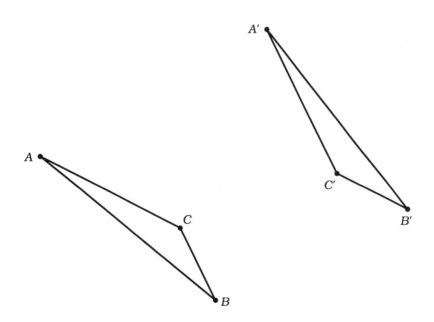

Figure T-65

SUPPLEMENTARY TOPICS 157

20. Develop an informal proof for Theorem T-13.

21. Develop an informal proof for Theorem T-14.

22. Develop an informal proof for Theorem T-15.

23. Develop an informal proof for Theorem T-16.

24. Develop an informal proof for Theorem T-17.

25. Develop an informal proof for Theorem T-18.

Chapter 7: Parallel and Perpendicular Lines; Additional Topics Concerning Angles, Triangles, and Segments

7.1 Introduction

The study of the geometry of a plane is concluded with an examination of two special relationships which exist among certain lines. The concept of parallel lines will be studied initially. This concept will be used to develop a collection of theorems involving angles, triangles, and segments. The second concept to be studied will be that of perpendicular lines. This concept will also be used to develop a collection of theorems involving angles, triangles, and segments.

7.2 Parallel Lines

This section contains several theorems pertaining to conditions which cause lines to be parallel and the consequences of lines being parallel. Some definitions will be helpful in describing certain situations.

Definition 27: Two lines which do not intersect are called **parallel lines**.

Language Rule 18: Two halflines, two segments, or a halfline and a segment which lie on parallel lines will be called **parallel**.

Definition 28: A line which has in common with two other lines exactly two points, one on each of the two lines, is called a **transversal** for the two lines.

Definition 29: Suppose m and n are two lines for which line k is a transversal, intersecting them at points A and B, respectively. Each of the two angles which has A for a vertex, a halfline of m for one side, and for its other side the halfline of k containing segment (A,B) is called an **interior angle**. Similarly, there are two interior angles having B as vertex. (See **Figure 7.1**.) Two interior angles, one having vertex A and the other having vertex B, whose interiors lie on opposite sides of k are called **alternate interior angles**.

Each of the two angles which has A for a vertex, a halfline of m for one side, and for its other side the halfline of k which does not contain (A,B) is called an **exterior angle**. Similarly, there are two exterior angles having B

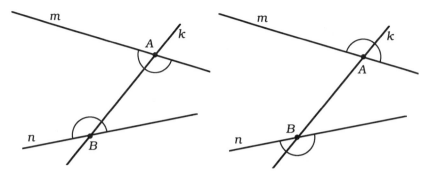

Figure 7.1   Figure 7.2

as vertex. (See **Figure 7.2**.)

Two angles such that one has vertex A and the other vertex B, one is an interior angle and the other is an exterior angle, and their interiors are on the same side of the transversal are called **corresponding angles**.

Theorem 76: If two lines intersected by a transversal have a pair of alternate interior angles which are congruent, then the two lines are parallel.

Proof: Suppose m and n are two lines intersected by a transversal at points A and B, respectively. Suppose, further, that these three lines form a congruent pair of alternate interior angles $\angle ABC$ and $\angle BAD$, where C and D are points on n and m, respectively. Finally, suppose that m and n are not parallel. (See **Figure 7.3**.)

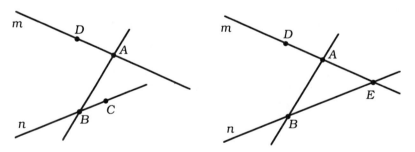

Figure 7.3

By Definition 29, C and D are on opposite sides of the transversal.

Since m and n are not parallel, then, by Definition 27, they intersect at a point E. Without loss of generality, assume that E is on the C-side of the transversal. (Other than the fact that E is on the C-side of B on line n, the relationship between C and E is irrelevant.) (See **Figure 7.3.**)

By Axiom 12, there exists on halfline $\overrightarrow{AD}$ a point F such that $(A,F) \cong (B,E)$. By Axiom 13, $(A,B) \cong (B,A)$. (See **Figure 7.4.**)

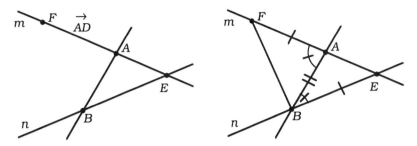

Figure 7.4

Since F is on $\overrightarrow{AD}$ then, by Theorem 21, F is on the D-side of the transversal, and, thus, F and E are on opposite sides of the transversal, by Theorem 22. Hence, A, B, F are distinct, noncollinear, and are the vertices of $\triangle ABF$.

Since A, B, E are distinct and noncollinear, they are the vertices of $\triangle ABE$.

In $\triangle ABF$ and $\triangle BAE$, $(A,F) \cong (B,E)$, $\angle BAF \cong \angle ABE$, and $(A,B) \cong (B,A)$. Hence, $\triangle ABF \cong \triangle BAE$, by Theorem 50. Thus, $\angle BAE \cong \angle ABF$, by Definition 16.

By Definition 22, $\angle BAF$ and $\angle BAE$ are supplementary angles.

Since $\angle ABE \cong \angle BAF$, then, by Theorem 66, the angle which is supplementary to $\angle ABE$ must be congruent to $\angle BAE$. By Axiom 15, the angle supplementary to $\angle ABE$ must be $\angle ABF$.

Thus, point F must be on line n.

Hence, Theorem 2 is contradicted since the two lines m and n have in common the two points E and F.

Therefore, the initial supposition that m and n are not parallel is false.

Thus, m and n are parallel. ∎

Theorem 77: If two lines intersected by a transversal have two interior angles which are supplementary and have their interiors on the same side of the transversal, then the two lines are parallel.

Theorem 78: If two lines intersected by a transversal have a pair of corresponding angles which are congruent, then the two lines are parallel.

The preceding three theorems have converses which are also theorems. However, before their inverses and two other theorems can be proved, it is necessary that an additional axiom be stated.

Axiom 18: If m is a line and A is a point which is not on m, then there exists exactly one line which contains A and is parallel to m.

Theorem 79: If two parallel lines are intersected by a transversal, then each pair of alternate interior angles are congruent.

Theorem 80: If two parallel lines are intersected by a transversal, then each pair of interior angles with their interiors on the same side of the transversal are supplementary.

Theorem 81: If two parallel lines are intersected by a transversal, then each pair of corresponding angles are congruent.

Theorem 82: If a line intersects one of two parallel lines, then it intersects the other.

Theorem 83: If m, n, t are three lines such that each of m and n is parallel to t, then m and n are parallel.

Exercises 7.2

1. Develop a proof for Theorem 77.

2. Develop a proof for Theorem 78.

3. Develop a proof for Theorem 79.

4. Develop a proof for Theorem 80.

5. Develop a proof for Theorem 81.

6. Develop a proof for Theorem 82.

7. Develop a proof for Theorem 83.

7.3 Interior Angles, Exterior Angles, Midpoints, and Bisectors

This section contains some definitions and theorems pertaining to triangles, angles, and segments. Although these might appear to have been misplaced, that is not the case. These theorems could not be presented earlier because their proofs are dependent on some of the material in the preceding section. In particular, they are dependent on materials which precede Axiom 18.

Definition 30: An angle which is adjacent and supplementary to an angle of a triangle is called an **exterior angle** of the triangle.

Language Rule 19: An angle of a triangle is sometimes called an **interior angle** of the triangle.

Definition 31: An exterior angle of a triangle which is not adjacent to an interior angle of the triangle is called a **remote exterior angle** of the interior angle.

Definition 32: An angle of a triangle which is not adjacent to an exterior angle of the triangle is called a **remote interior angle** of the exterior angle.

Theorem 84: An interior angle of a triangle is less than each of its remote exterior angles.

Proof: Suppose △ABC; without loss of generality, consider ∠BAC.

By Axiom 5, Axiom 6, and Axiom 1b, there exist points D and E on sidelines AC and BC, respectively, such that A•C•D and B•C•E. Analogously, there exist points F and G on sidelines BC and AB, respectively, such that F•B•C and A•B•G. (See **Figure 7.5**.)

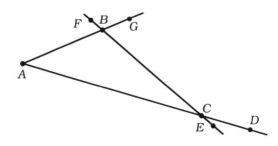

Figure 7.5

By Definition 21, Definition 30, and Definition 31, each of ∠ABF, ∠CBG, ∠BCD, and ∠ACE is a remote exterior angle of interior angle ∠BAC.

Claim: ∠BAC < ∠ABF.

Proof: By Theorem 64, exactly one of ∠BAC < ∠ABF, ∠BAC ≅ ∠ABF, and ∠ABF < ∠BAC is true.

Subclaim: ∠BAC ≅ ∠ABF is not true.

Proof: Suppose ∠BAC ≅ ∠ABF.
By Definition 28, line AB is a transversal for lines BC and AC. By Definition 29, ∠BAC and ∠ABF are alternate interior angles for lines BC and AC and transversal AB.
Since AC and BC intersect at C, then, by Definition 27, they are not parallel. Hence, Theorem 76 is contradicted.
Thus, the supposition is false.
Therefore, ∠BAC ≆ ∠ABF. ♦♦

Subclaim: ∠ABF < ∠BAC is not true.

Proof: Suppose ∠ABF < ∠BAC.
By Definition 19, there exists a halfline $\overrightarrow{AH}$ such that $\overrightarrow{AH}$ is between $\overrightarrow{AB}$ and $\overrightarrow{AC}$ and ∠ABF ≅ ∠BAH. (See **Figure 7.6**.)

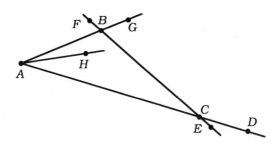

Figure 7.6

By Definition 15 and Theorem 41, $\overrightarrow{AH}$ intersects side (B,C) at a point J. (See **Figure 7.7**.)

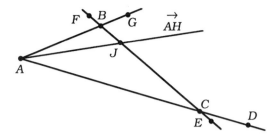

Figure 7.7

Since ∠ABF ≅ ∠BAH and ∠BAH = ∠BAJ, then ∠ABF ≅ ∠BAJ, by Axiom 16.

By Definition 28, line AB is a transversal for lines BJ and AJ. By Definition 29, ∠ABF and ∠BAJ are alternate interior angles for lines BJ and AJ and transversal AB.

Since AJ and BJ intersect at J, then, by Definition 27, they are not parallel. Hence, Theorem 76 is contradicted.

Thus, the supposition is false.

Therefore, ∠ABF ≮ ∠BAC. ♦♦

Thus, ∠BAC ≇ ∠ABF and ∠ABF ≮ ∠BAC. Therefore, ∠BAC < ∠ABF. ♦

Claim: ∠BAC < ∠ACE.

Proof: Analogous to the preceding **Claim**. ♦

Claim: ∠BAC < ∠CBG.

Proof: Left as an exercise. ♦

Claim: ∠BAC < ∠BCD.

Proof: Analogous to the preceding **Claim**. ♦

Therefore, an interior angle of a triangle is less than each of its remote exterior angles. ∎

Theorem 85: If two sides of a triangle are not congruent, then their opposite angles are not congruent and the lesser angle is opposite the lesser side.

Theorem 86: If two angles of a triangle are not congruent, then their opposite sides are not congruent and the lesser side is opposite the lesser angle.

With the use of Theorem 84, it is possible to establish a fourth triangle congruence theorem which can be used in a manner similar to the three triangle congruence theorems which were established in Chapter 6.

Theorem 87: If $\triangle ABC$ and $\triangle DEF$ are triangles such that $(A,B) \cong (D,E)$, $\angle A \cong \angle D$, and $\angle C \cong \angle F$, then $\triangle ABC \cong \triangle DEF$.

Proof: Suppose $\triangle ABC$ and $\triangle DEF$ are triangles such that $(A,B) \cong (D,E)$, $\angle A \cong \angle D$, and $\angle C \cong \angle F$. (See **Figure 7.8**.)

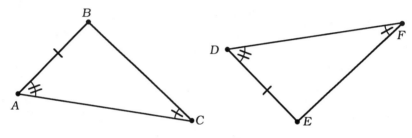

Figure 7.8

By Theorem 60, exactly one of $(A,C) \cong (D,F)$, $(A,C) < (D,F)$, and $(D,F) < (A,C)$ is true.

Claim: $(A,C) < (D,F)$ is not true.

Proof: Left as an exercise. ◆

Claim: $(D,F) < (A,C)$ is not true.

Proof: Analogous to the preceding **Claim**. ◆

Thus, $(A,C) \not< (D,F)$ and $(D,F) \not< (A,C)$. Therefore, $(A,C) \cong (D,F)$.

Since $(A,C) \cong (D,F)$, then, in $\triangle ABC$ and $\triangle DEF$, $(A,B) \cong (D,E)$, $\angle A \cong \angle D$, and $(A,C) \cong (D,F)$. By Theorem 50, $\triangle ABC \cong \triangle DEF$. ∎

Language Rule 20: The preceding theorem can be called the **Side**

Angle Angle congruence theorem and restated as "If two angles and a nonincluded side of one triangle are congruent, respectively, to two angles and a nonincluded side of another triangle, then the triangles are congruent."

Definition 33: If C is a point on a segment (A,B) such that $(A,C) \cong (C,B)$, then C is called a **midpoint** of the segment.

Definition 34: If a line m intersects a segment (A,B) at its midpoint C, then m is called a **bisector** of (A,B).

Theorem 88: Every segment has exactly one midpoint.

Proof: Suppose (A,B) is a segment.

By Axiom 3, Axiom 1a, Definition 2, and Definition 12, there exists an angle $\angle ABC$.

By Axiom 15 and Axiom 12, there exists a halfline $\overrightarrow{AD}$ on the opposite side of line AB from $\overrightarrow{BC}$ such that $\angle BAD \cong \angle ABC$ and $(A,D) \cong (B,C)$. (See **Figure 7.9**.)

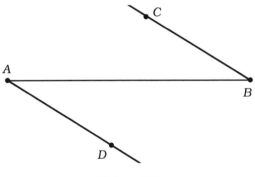

Figure 7.9

Since $\overrightarrow{BC}$ and $\overrightarrow{AD}$ are on opposite sides of AB, then, by Theorem 21 and Theorem 22, C and D are on opposite sides of AB.

Claim: (A,B) has at least one midpoint.

Proof: By Theorem 20, there exists a point E such that E is on AB and

$C \bullet E \bullet D$. By Axiom 6, C, E, D are collinear.

There are two possibilities: $E \in \{A,B\}$ or $E \notin \{A,B\}$.

Subclaim: $E \in \{A,B\}$ is not true.

Proof: Left as an exercise. ♦♦

Since E is on AB and $E \notin \{A,B\}$, then, by Axiom 8, exactly one of $E \bullet A \bullet B$, $A \bullet E \bullet B$, and $A \bullet B \bullet E$ is true.

Subclaim: $E \bullet A \bullet B$ is not true.

Proof: Left as an exercise. ♦♦

Subclaim: $A \bullet B \bullet E$ is not true.

Proof: Analogous to the preceding **Claim.** ♦♦

Since each of $E \bullet A \bullet B$ and $A \bullet B \bullet E$ is not true, then $A \bullet E \bullet B$. (See **Figure 7.10**.)

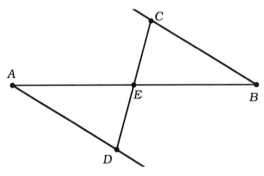

Figure 7.10

Thus, $\overrightarrow{EC}$, $\overrightarrow{ED}$ are opposite halflines, and $\overrightarrow{EA}$, $\overrightarrow{EB}$ are opposite halflines.

By Definition 23, $\angle AED$ and $\angle BEC$ are vertical angles. Hence, by Theorem 68, $\angle AED \cong \angle BEC$.

Thus, in $\triangle AED$ and $\triangle BEC$, $(A,D) \cong (B,C)$, $\angle EAD \cong \angle EBC$, and $\angle AED \cong \angle BEC$. By Theorem 87, $\triangle AED \cong \triangle BEC$. Hence, by Definition 16, $(A,E) \cong (B,E)$.

Therefore, by Definition 33, E is a midpoint of (A,B). ♦

Claim: (A,B) has at most one midpoint.

Proof: Left as an exercise. ♦

Therefore, (A,B) has exactly one midpoint. ∎

Definition 35: If halfline $\overrightarrow{OC}$ is between halflines $\overrightarrow{OA}$ and $\overrightarrow{OB}$ such that $\angle AOC \cong \angle BOC$, then $\overrightarrow{OC}$ is called a **bisector** of $\angle AOB$.

Theorem 89: Every angle has exactly one bisector.

Exercises 7.3

1. Complete the proof for Theorem 84.
2. Develop a proof for Theorem 85.
3. Develop a proof for Theorem 86.
4. Complete the proof for Theorem 87.
5. Complete the proof for Theorem 88.
6. Develop a proof for Theorem 89.

7.4 Perpendicular Lines

This section contains two theorems concerning the existence of perpendicular lines. In addition, there are three theorems relating this concept to previously examined concepts.

Definition 36: Two noncollinear halflines with a common endpoint are called **perpendicular** if the angle they form with their endpoint is a right angle.

Language Rule 21: The common endpoint of perpendicular halflines will be called the **point of perpendicularity**. Also, the halflines are said to be **perpendicular** to each other.

Definition 37: Two intersecting lines which contain perpendicular halflines are called **perpendicular lines**.

Language Rule 22: Two segments which lie on perpendicular lines and intersect or have a common endpoint will be called **perpendicular segments**. A line and a segment will be called **perpendicular line and segment** if the segment intersects the line or has an endpoint on the line and lies on a second line which is perpendicular to the first line. (See **Figure 7.11**.)

Figure 7.11

Notation Rule 17: Perpendicularity will be indicated in diagrams with the use of a small square, as illustrated in **Figure 7.11**.

Language Rule 23: Two perpendicular lines, two perpendicular segments, or perpendicular line and segment are said to be **perpendicular** to each other.

Theorem 90: If m is a line and A is a point not on m, then there exists exactly one line which contains A and is perpendicular to m.

Proof: Suppose m is a line and A is a point not on m.

Claim: There exists at least one line which contains A and is perpendicular to m.

Proof: By Axiom 2, there exist at least two points B and C on m.
Since A is not on m and B, C are on m, then A is distinct from each of B and C. By Axiom 1, there exists a line AB containing A and B. By Definition 2, halfline $\overrightarrow{BA}$ lies on AB and halfline $\overrightarrow{BC}$ lies on m. Since A is not on m, $\overrightarrow{BA}$ and $\overrightarrow{BC}$ are noncollinear.

By Axiom 15, there exists a halfline $\overrightarrow{BD}$ on the opposite side of m from $\overrightarrow{BA}$ such that $\angle ABC \cong \angle DBC$. (See **Figure 7.12**.)

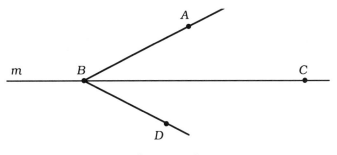

Figure 7.12

By Axiom 12, there exists on $\overrightarrow{BD}$ a point E such that $(B,A) \cong (B,E)$.
Since $\overrightarrow{BA}$ and $\overrightarrow{BD}$ lie on opposite sides of m and contain A and E, respectively, then, by Theorem 21, A and E are on opposite sides of m.

Since A and E are distinct, by Axiom 1, they determine line AE. By Theorem 20 and Definition 4, AE intersects m at a point F such that $A \bullet F \bullet E$. (See **Figure 7.13**.) Thus, $\overrightarrow{FA}$ and $\overrightarrow{FE}$ are opposite halflines.

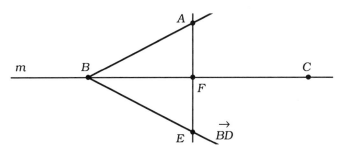

Figure 7.13

There are two possibilities: $F = B$ or $F \neq B$.

Subclaim: If $F = B$, then line AB is perpendicular to m.

Proof: Suppose $F = B$.

By Definition 21, $\angle AFC$ and $\angle EFC$ are a linear pair of angles. Hence, by Axiom 16 and Definition 22, $\angle AFC$ and $\angle EFC$ are supplementary angles.

Since $F = B$, then $\angle ABC$ and $\angle EBC$ are supplementary angles. Since E is a point of $\overrightarrow{BD}$, then $\overrightarrow{BD} = \overrightarrow{BE}$. Thus, $\angle ABC$ and $\angle DBC$ are supplementary angles.

By Definition 24, ∠ABC is a right angle since it is congruent to its supplementary angle.

Hence, by Definition 36, halflines $\overrightarrow{BA}$ and $\overrightarrow{BC}$ are perpendicular. Since $\overrightarrow{BA}$ lies on AB and $\overrightarrow{BC}$ lies on m, then, by Definition 37, AB is perpendicular to m. ♦♦

Subclaim: If $F \neq B$, then line AF is perpendicular to m.

Proof: Left as an exercise. ♦♦

Therefore, if m is a line and A is a point not on m, then there exists at least one line which contains A and is perpendicular to m. ♦

Claim: There exists at most one line which contains A and is perpendicular to m.

Proof: Left as an exercise. ♦

Therefore, if m is a line and A is a point not on m, then there exists exactly one line which contains A and is perpendicular to m. ■

Theorem 91: There exists at least one right angle.

Theorem 92: If m is a line and A is a point on m, then there exists exactly one line which contains A and is perpendicular to m.

Theorem 93: If a line is perpendicular to one of two parallel lines, then it is perpendicular to the other.

Theorem 94: If two lines are perpendicular, respectively, to a pair of parallel lines, then the lines are parallel.

Exercises 7.4

1. Complete the proof for Theorem 90.

2. Develop a proof for Theorem 91.

3. Develop a proof for Theorem 92.

4. Develop a proof for Theorem 93.

5. Develop a proof for Theorem 94.

7.5 Some Additional Theorems Concerning Angles, Triangles, and Segments

This section contains some definitions and theorems pertaining to angles, triangles, and segments. Although these might appear to have been misplaced, that is not the case. These theorems could not be presented earlier because their proofs are dependent on some of the material in the preceding section.

Theorem 95: If m is a line, A is a point not on m, B is the point on m such that (A,B) is perpendicular to m, and C is a second point on m, then $(A,B) < (A,C)$.

Theorem 96: At least two angles of any triangle are acute.

Proof: Suppose $\triangle ABC$; without loss of generality, consider exterior angle $\angle DBC$. (See **Figure 7.14**.)

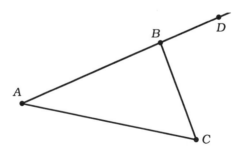

Figure 7.14

There exists at least one right angle $\angle R$, by Theorem 91.
By Theorem 64, exactly one of $\angle DBC < \angle R$, $\angle DBC \cong \angle R$, and $\angle R < \angle DBC$ is true.

Claim: If $\angle DBC < \angle R$, then $\triangle ABC$ has at least two acute angles.

Proof: Left as an exercise. ♦

Claim: If ∠DBC ≅ ∠R, then △ABC has at least two acute angles.

Proof: Left as an exercise. ♦

Claim: If ∠R < ∠DBC, then △ABC has at least two acute angles.

Proof: Left as an exercise. ♦

Therefore, △ABC has at least two acute angles. ■

Definition 38: A line which is perpendicular to a segment at its midpoint is called a **perpendicular bisector** of the segment.

Theorem 97: Every segment has exactly one perpendicular bisector.

Definition 39: If A, B, C are three points such $(A,C) \cong (C,B)$, then C is said to be **equidistant** from A and B.

Theorem 98: A point is equidistant from the endpoints of a segment if and only if it lies on the perpendicular bisector of the segment.

Exercises 7.5

1. Develop a proof for Theorem 95.

2. Complete the proof of Theorem 96.

3. Develop a proof for Theorem 97.

4. Develop a proof for Theorem 98.

5. Attempt to develop a proof for the following statement; if you cannot do so, explain the difficulty involved.
   A point is equidistant from the sides of an angle if and only if it lies on the bisector of the angle.

****

SUPPLEMENTARY TOPICS

CONSTRUCTIONS

29. Construct a triangle $\triangle ABC$ such that side $\overline{AB}$ is congruent to the given segment $\overline{DE}$, the median to side $\overline{BC}$ is congruent to the given segment $\overline{FG}$, and the median to side $\overline{AC}$ is congruent to the given segment $\overline{HJ}$.

```
_____   _____   _____
D           E    F          G   H                     J
```

30. Construct a triangle $\triangle ABC$ such that $\angle A$ is congruent to the given $\angle D$, side $\overline{AB}$ is congruent to the given segment $\overline{EF}$, and the portion of the bisector of $\angle A$ which is between A and side $\overline{BC}$ is congruent to the given segment $\overline{GH}$.

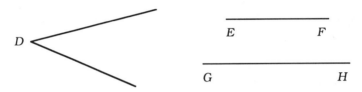

31. Using an initial segment measuring 1 inch, select any three noncollinear points A, B, C, such that they are located between 2 and 3 inches apart. Determine all points which are located 1.5 inches from segment $\overline{AB}$ and also $\sqrt{3}$ inches from point C.

32. Construct a triangle $\triangle ABC$ such that side $\overline{AB}$ is congruent to the given segment $\overline{DE}$, the median to side $\overline{AB}$ is congruent to the given segment $\overline{FG}$, and the median to side $\overline{BC}$ is congruent to the given segment $\overline{HJ}$.

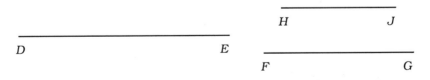

33. Select any four points A, B, C, D, such that no three of them are collinear and they do not lie together in pairs on parallel lines. Using an initial segment measuring 1 inch, determine all points which are located at the distance $\sqrt{5}$ from the point which is both equidistant from A and B and equidistant from C and D.

*****

PROBLEMS

15. In the diagram below, $\overline{AE}$ and $\overline{CD}$ are medians of $\triangle ABC$, G is the midpoint of $\overline{AF}$, H is the midpoint of $\overline{CF}$, $m\overline{AC} = 8$ units, $m\overline{AE} = 6$ units, and $m\overline{CD} = 6.6$ units. Determine the area and perimeter of the region enclosed by $\triangle DEF$.

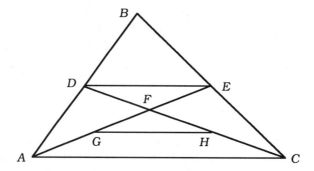

16. Show that the bisector of the right angle of a right triangle bisects the angle included between the altitude and the median on the hypotenuse.

*****

TRANSFORMATION GEOMETRY

T-7: Symmetry

Definition T-13: A set S of points has **line symmetry** if there exists a line m such that $\mathbf{R}_m(S) = S$. Line m is called a **line of symmetry**.

Remark T-7: If m is a **line of symmetry** for a set S of points, then $\mathbf{R}_m$ fixes S.

Example T-22: Dashed lines are used to indicate all the lines of symmetry in the diagrams in **Figure T-66**. ■

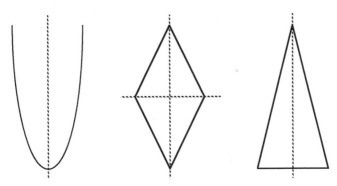

Figure T-66

Definition T-14: A set S of points has **rotational symmetry** if there exists a point P such that $\mathbf{P}_x(S) = S$, where $0° \leq x < 360°$.

Remark T-8: If P is a point for which $\mathbf{P}_x(S) = S$ for a set S of points, then $\mathbf{P}_x$ fixes S.

Definition T-15: A rotational symmetry with $x = 0°$ is called the **identity symmetry**.

Example T-23: In **Figure T-67**, the diagram on the left has rotational symmetries of $\mathbf{P}_0$, $\mathbf{P}_{120}$, $\mathbf{P}_{240}$, while the diagram on the right has rotational symmetries of $\mathbf{P}_0$, $\mathbf{P}_{90}$, $\mathbf{P}_{180}$, $\mathbf{P}_{270}$. ■

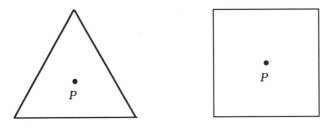

Figure T-67

Definition T-16: A set S of points has **point symmetry** if there exists a point P such that $\mathbf{P}_{180}(S) = S$. Point P is called a **point of symmetry**.

Remark T-9: If P is a **point of symmetry** for a set S of points, then $\mathbf{P}_{180}$ **fixes** S.

Example T-24: In each diagram in **Figure T-68**, the point of symmetry is indicated. ∎

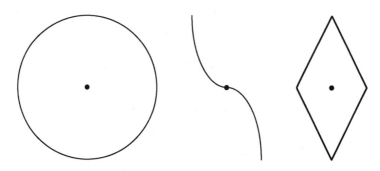

Figure T-68

Definition T-17: An isometry **M** is called a **symmetry** for a set S of points if $\mathbf{M}(S) = S$.

Remark T-10: If an isometry **M** is a **symmetry** for a set S of points, then **M fixes** S.

Example T-25: The diagram in **Figure T-69** has six line symmetries, as indicated by dashed lines. It also has six rotational symmetries: $\mathbf{P}_0$, $\mathbf{P}_{60}$, $\mathbf{P}_{120}$, $\mathbf{P}_{180}$, $\mathbf{P}_{240}$, $\mathbf{P}_{300}$, with $\mathbf{P}_0$ being the identity symmetry and $\mathbf{P}_{180}$ being a point symmetry.

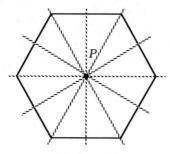

Figure T-69

Exercises T-7

1. For each diagram in **Figure T-70**, identify *all* its symmetries.

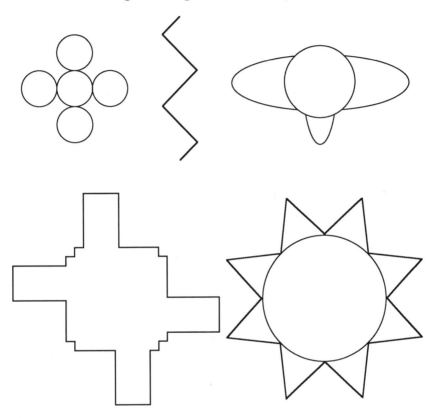

Figure T-70

2. If *S* is the set of points in a regular polygon having *n* sides, identify *all* the symmetries for *S*.

3. For each symbol in **Figure T-71**, identify *all* its symmetries.

$$\Pi \quad \S \quad \text{¤} \quad \ddagger \quad 8 \quad \Xi \quad \forall \quad \exists \quad \perp \quad \uparrow$$
$$\leftrightarrow \quad \int \quad \in \quad \pm \quad \cong \quad \# \quad \emptyset \quad \Omega \quad \infty \quad [$$

Figure T-71

4. For each symbol in **Figure T-72**, identify *all* its symmetries.

Figure T-72

Chapter 8: The Foundations of Space Geometry

8.1 Introduction

While the material in the preceding four chapters provided a study of some subsets of a plane, the material in this chapter will involve a brief examination of some axioms and theorems pertaining to the geometry of space, with the primary emphasis being placed on existence and incidence. As was indicated in the introduction to Chapter 4, the material presented in that chapter can be used in proving theorems in this chapter, since every plane and its subsets will possess the same characteristics as those established for the one plane considered in Chapter 4.

In order to facilitate the transition from the material in Chapter 4 to the related material in this chapter, some of the definitions, axioms, theorems, and language and notation rules from Chapter 4 will be restated in the appropriate locations in this chapter. Although they will be stated with their original numbers, they may contain additions pertaining to space. In all cases, these additions will be italicized.

8.2 Existence and Incidence

The axioms and theorems of this section establish connections between the existence and incidence of points, lines, and planes.

Language Rule 1: Point, **line**, and **plane**, and the vocabulary of logic and sets will be undefined terms.

Language Rule 2: Points will be considered the elements of **linear geometry**; points and lines will be considered the elements of **plane geometry**; *points, lines, and planes will be considered the elements of **space geometry**.*

Notation Rule 1: Points will be denoted with upper case letters, lines will be denoted with lower case letters, and planes will be denoted with boldface upper case letters.

Language Rule 4: Statements such as *m* **contains** *A*, *A* **lies on** *m*, and *m* **passes through** *A* will be used to indicate that a point *A* is an element of a line *m*. *Analogous statements will be used for a point and a plane and for a line and a plane.*

Language Rule 5: Statements such as **m and n have A in common** and **m and n intersect at A** will be used to indicate that a point A is an element of two lines. Similar statements will be used when a point A is an element of more than two lines. *Analogous statements will be used for a point and planes and for a line and planes.*

Axiom 1a: If A and B are two points, then there exists at least one line containing them.

Axiom 1b: If A and B are two points, then there exists at most one line containing them.

Axiom 2: If m is a line, then it contains at least two points.

Axiom 3: If m is a line, then there exists at least one point which does not lie on it.

Axiom 4: There exist at least two points.

Theorem 1: There exist at least three points.

Theorem 2: Two lines have at most one point in common.

Definition 1: Two or more points are called **collinear** if they lie on the same line.

Theorem 3: If A and B are two points on a line m and C is a point not on m, then A, B, C are noncollinear and there exist lines n and k, containing A, C and B, C, respectively, which are distinct from m and from each other.

Definition 40: Two or more points are called **coplanar points** if they lie on the same plane. Two or more lines are called **coplanar lines** if they lie on the same plane.

Theorem 4: There exist at least two *coplanar* lines through any point.

Theorem 5: If A is a point *on a plane P*, then there exists at least one line *on P* which does not contain it.

Axiom 19a: If A, B, C are three noncollinear points, then there exists at least one plane containing them.

Axiom 19b: If *A*, *B*, *C* are three noncollinear points, then there exists at most one plane containing them.

Notation Rule 18: If *A*, *B*, *C* are any three noncollinear points on a plane ***P***, then ***P*** may also be denoted as *ABC*.

Axiom 20: If ***P*** is a plane, then it contains at least three noncollinear points.

Axiom 21: If at least two points of a line *m* lie on a plane ***P***, then every point lying on *m* lies on ***P***.

Axiom 22: If two planes have a point in common, then they have at least one more point in common.

Axiom 23: If ***P*** is a plane, then there exists at least one point which does not lie on it.

Definition 41: Four or more points are called **space points** if they are not necessarily coplanar. Two or more lines are called **space lines** if they are not necessarily coplanar.

Theorem 99: If *m* is a line and *A* is a point not on *m*, then there exists exactly one plane containing *m* and *A*.

Theorem 100: If *m* and *n* are two lines which have a point in common, then there exists exactly one plane containing them.

Theorem 101: Two lines lie together on at most one plane.

Theorem 102: Not all lines lie on the same plane.

Theorem 103: If *A* is a point on a plane ***P***, then there exists at least one line which does not contain *A* and does not lie on ***P***.

Theorem 104: There exist at least two planes through any point.

Theorem 105: If *A* is a point, then there exists at least one plane which does not contain it.

Theorem 106: If two planes have a point in common, then all their common points lie on exactly one line.

Theorem 107: If ***P*** is a plane and *m* is a line which does not lie on ***P***, then ***P*** and *m* have at most one point in common.

Theorem 108: There exist at least six space lines.

Theorem 109: There exist at least four planes.

Exercises 8.2

1. Develop a proof for Theorem 99.
2. Develop a proof for Theorem 100.
3. Develop a proof for Theorem 101.
4. Develop a proof for Theorem 102.
5. Develop a proof for Theorem 103.
6. Develop a proof for Theorem 104.
7. Develop a proof for Theorem 105.
8. Develop a proof for Theorem 106.
9. Develop a proof for Theorem 107.
10. Develop a proof for Theorem 108.
11. Develop a proof for Theorem 109.

8.3 Halfspaces; Partition; Separation; Convexity

Definition 42: If ***P*** is a plane and *A* is a point which is not on ***P***, then
$$S_1 = \{A\} \cup \{X \mid X \in \boldsymbol{P} \text{ and } (A,X) \cap \boldsymbol{P} = \varnothing\}$$
$$\text{and } S_2 = \{X \mid X \in \boldsymbol{P} \text{ and } (A,X) \cap \boldsymbol{P} \neq \varnothing\}$$
are called the **halfspaces** with respect to ***P*** relative to *A*.

Language Rule 24: The halfspaces based on a plane ***P*** can be described as having been determined by **P** and will be called the **sides of space relative to P**. Two points lying on the same halfspace will be described as being **on the same side of P**, while two points lying on different halfspaces will be described as being on **opposite sides of P**.

Definition 3: If a set S can be expressed as the union of two or more nonempty, disjoint sets S_1, S_2, ..., then S_1, S_2, ... are said to **partition** S, and the sets S_1, S_2, ... are called a **partition** of S.

Theorem 110: If **P** is a plane and A is a point which is not on **P**, then **P** and S_1, S_2, the halfspaces with respect to **P** relative to A, partition space.

Definition 4: If S, S_1, S_2 are three nonempty, disjoint sets for which it is true that
(1)   if X is an element of S_1 and Y is an element of S_2, then there exists an element Z of S such that $X \bullet Z \bullet Y$; and
(2)   if X, Y are two elements of S_1, then there does not exist an element Z of S such that $X \bullet Z \bullet Y$; and
(3)   if X, Y are two elements of S_2, then there does not exist an element Z of S such that $X \bullet Z \bullet Y$;
then S is said to **separate** S_1 and S_2.

Theorem 111: If **P** is a plane, then it separates the two halfspaces which it determines.

Definition 7: A set S of points is called **convex** if, for any two points A, B belonging to S, the segment (A,B) is a subset of S.

Theorem 112: Every halfspace, with respect to a plane **P**, is convex.

Exercises 8.3

1. Develop a proof for Theorem 110.

2. Develop a proof for Theorem 111.

3. Develop a proof for Theorem 112.

8.4 Some Additional Theorems Concerning Existence and Incidence

Theorem 25: There exist infinitely many distinct points on any line.

Theorem 113: There exist infinitely many distinct lines on any plane.

Theorem 26: There exist infinitely many distinct *coplanar* lines through any point.

Theorem 27: If A is a point *on a plane P*, then there exist infinitely many distinct lines *on P* which do not contain A.

Theorem 114: If A is a point on a plane P, then there exist infinitely many distinct space lines which do not contain A and do not lie on P.

Theorem 115: There exist infinitely many distinct planes through any point.

Exercises 8.4

1. Develop a proof for Theorem 113.

2. Develop a proof for Theorem 114.

3. Develop a proof for Theorem 115.

****

SUPPLEMENTARY TOPICS

CONSTRUCTIONS

34. Construct a segment $\overline{AB}$ congruent to the given segment $\overline{DE}$. Choose any point which is not on line AB and label it as F. Construct $\triangle ABC$ such that F will be the centroid of $\triangle ABC$.

$$D \qquad\qquad\qquad E$$

35. Construct two parallel lines which are located about 2 inches apart,

and select a point A which is located anywhere except on or between the lines. Construct a line m which passes through A in such a manner that the two parallel lines cut off on m a segment $\overline{BC}$, congruent to the given segment $\overline{DE}$.

D   E
---

36. Construct a circle, and draw a line which is located about 1 inch from the circle. Construct a circle which has its center on the line, has the measure of its radius equal to the measure of the given segment $\overline{AB}$, and is tangent to the initial circle. How many such circles are possible?

A   B
---

37. Construct two nonintersecting, noncongruent circles such that neither is inside the other, and denote their centers as C and D. Determine two points, E and F, one on each of the given circles and on the same side of line CD such that line EF will be tangent to each of the given circles.

*****

PROBLEMS

17. Some isosceles triangular regions can be partitioned into other noncongruent, disjoint isosceles triangular regions with the addition of certain line segments. Calculations will show that not every isosceles triangular region can be partitioned in this manner. In fact, the vertex angle must be a certain size for each case; for example, an isosceles triangular region can be partitioned into two such triangular regions only when its vertex angle has measure 36°. Determine a formula which will show the measure of the vertex angle of an isosceles triangular region which can be partitioned into n noncongruent, disjoint isosceles triangular regions. (This problem is based on the following article: Isosceles^n, E. R. Ranucci, *Mathematics Teacher*, April, 1976.)

18. In a large, level city park, four rest areas are located at the vertices of a square of side 1 mile. The city recreation department wishes to install a system of walking paths to connect the rest areas, but

they have only enough materials and funds to make $1+\sqrt{3}$ miles of path. What must be the configuration of the paths, if there is to be exactly $1+\sqrt{3}$ miles of path?

*****

TRANSFORMATION GEOMETRY

T-8: Similarity Transformations

Our study of transformation geometry concludes with an examination of a very important type of transformation which is not always an isometry.

Definition T-18: If **T** is a transformation of a plane and k is a positive number such that for any two points P and Q of the plane, $m\overline{P'Q'} = k(m\overline{PQ})$ where $\mathbf{T}(P) = P'$ and $\mathbf{T}(Q) = Q'$, then **T** is a **similarity**. The number k is the **ratio of similarity** for **T**.

Theorem T-19: A similarity preserves collinearity.

Theorem T-20: A similarity maps lines to lines, segments to segments, rays to rays, and angles to angles.

Theorem T-21: A similarity preserves betweenness and maps midpoints to midpoints.

Theorem T-22: A similarity preserves angle measure.

As the following definitions and theorems indicate, there are several transformations and compositions of transformations which may be classified as similarity transformations.

Definition T-19: If C is a point in a plane and k is a positive number, then a **stretch of ratio k from C** is a transformation of the plane such that

$$\mathbf{S}_{C,k}(P) = \begin{cases} P' = P, \text{ if } P = C \\ P', \text{ where } P \text{ is the point on } \overline{CP} \\ \text{such that } m\overline{CP'} = k(m\overline{CP}), \text{ if } P \neq C. \end{cases}$$

Remark T-11: Although no *stretching* occurs when $k = 1$ in a **stretch**, this transformation, which is the **identity** transformation, will be called a **stretch**.

Remark T-12: Although *shrinking* actually occurs when $0 < k < 1$ in a **stretch**, the transformation will be called a **stretch**.

Remark T-13: The point C is often called the **center** of the stretch.

Example T-26: When a polygon is stretched from a point, its image can be found by locating the images of its vertices and sketching the polygon which they determine. In **Figure T-73**, $k = 3$. ∎

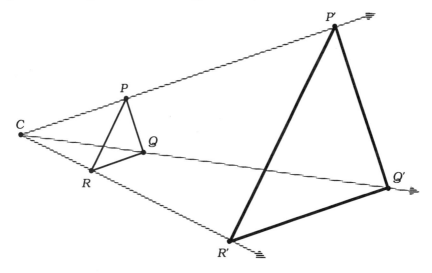

Figure T-73

Theorem T-23: A stretch is a similarity.

Example T-27: When a circle is stretched from a point, its image can be found by locating the image of its center and sketching the circle with this image point as center and having a radius which is k times the radius of the original circle. In **Figure T-74**, $k = 0.5$. ∎

Definition T-20: If C is a point in a plane, k is a positive number, and m is a line through C, then the transformations composition $T_{C,k,m} = R_m \circ S_{C,k}$ is a **stretch reflection from C across m**.

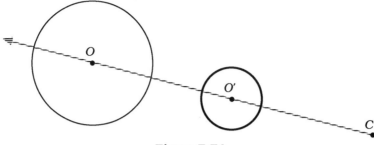

Figure T-74

Example T-28: In **Figure T-75**, C is the center of a stretch reflection **T** from C across m with $k = 2$. $\mathbf{S}_{C,2}(P) = P^*$ and $\mathbf{R}_m(P^*) = P'$. Hence, $\mathbf{T}_{C,2,m}(P) = \mathbf{R}_m(\mathbf{S}_{C,2}(P)) = \mathbf{R}_m(P^*) = P'$. ∎

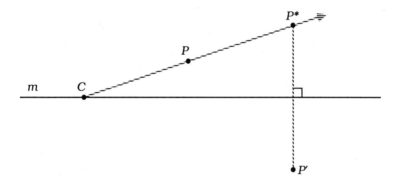

Figure T-75

Theorem T-24: A stretch reflection is a similarity.

Definition T-21: If C is a point in a plane, k is a positive number, and x is any number, then the transformations composition $\mathbf{T}_{C,k,x} = \mathbf{C}_x \circ \mathbf{S}_{C,k}$ is a **stretch rotation from C about C**.

Figure T-76

Example T-29: In **Figure T-76**, C is the center of a stretch rotation $\mathbf{T}$ from C about C with $k = 0.25$ and $x = 45°$. $\mathbf{S}_{C,0.25}(P) = P^*$ and $\mathbf{C}_{45}(P^*) = P'$. Hence, $\mathbf{T}_{C,0.25,45}(P) = \mathbf{C}_{45}(\mathbf{S}_{C,0.25}(P)) = \mathbf{C}_{45}(P^*) = P'$. ∎

Theorem T-25: A stretch rotation is a similarity.

Definition T-22: If C is a point in a plane and z is a nonzero number, then a **dilation from** C is a transformation of the plane such that

$$\mathbf{D}_{C,z} = \begin{cases} \mathbf{S}_{C,z}, & \text{if } z > 0 \\ \mathbf{C}_{180} \circ \mathbf{S}_{C,-z}, & \text{if } z < 0. \end{cases}$$

Theorem T-26: A dilation is a similarity.

This section concludes with a small collection of theorems describing the relationships between certain types of transformations.

Theorem T-27: An isometry is a similarity.

Theorem T-28: A similarity which fixes at least two points is an isometry.

Theorem T-29: A similarity which fixes at least three points is the identity transformation.

Theorem T-30: Every similarity is an isometry, a stretch, a stretch reflection, or a stretch rotation.

Exercises T-8

1. Does a similarity preserve parallelism? Why?

2. Does a similarity preserve perpendicularity? Why?

3. Suppose a similarity transformation $\mathbf{T}$, having similarity ratio k, is performed on the following objects. In each case, how will the area of the image's interior compare with area of the original object's interior? Why?
   (a) a rectangle
   (b) a triangle
   (c) a circle
   (d) a polygon

(e) any closed figure on a plane

4. Construct a triangle congruent to the one in **Figure T-77**. Construct a second triangle whose interior area will be ⅑ that of the first triangle.

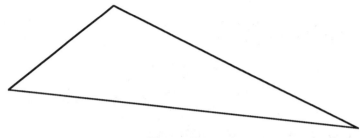

Figure T-77

5. If **T** is a similarity such that **T**(2,2) = (2,12) and **T**(1,–5) = (–1,–9), what is the similarity ratio for **T**?

6. If the similarity in problem 5 is a dilation, what is its center?

7. In **Figure T-78**, C is the center of a stretch reflection **T** from C across m. Construct a diagram identical to this one, and use constructions to determine the image of Q and the preimage of R'.

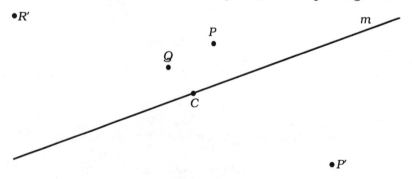

Figure T-78

8. In **Figure T-79**, C is the center of a stretch rotation **T** from C about C. Construct a diagram identical to this one, and use constructions to determine the image of Q and the preimage of R'.

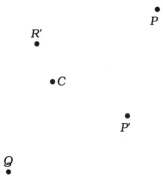

Figure T-79

9. In **Figure T-80**, C is the center of a dilation **D** from C. Construct a diagram identical to this one, and use constructions to determine the image of Q and the preimage of R'.

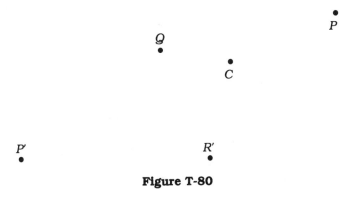

Figure T-80

10. In **Figure T-81**, the square has one side on side $\overline{AC}$ of $\triangle ABC$ and one vertex on side $\overline{AB}$. Construct a diagram identical to this one, and use constructions to determine a square which has one of its sides on $\overline{AC}$, one vertex on $\overline{AB}$, and one vertex on $\overline{BC}$. Give a proof to show that the quadrilateral you constructed is, in fact, a square.

11. Develop an informal proof for Theorem T-19.

12. Develop an informal proof for Theorem T-20.

13. Develop an informal proof for Theorem T-21.

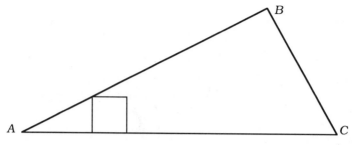

Figure T-81

14. Develop an informal proof for Theorem T-22.

15. Develop an informal proof for Theorem T-23.

16. Develop an informal proof for Theorem T-24.

17. Develop an informal proof for Theorem T-25.

18. Develop an informal proof for Theorem T-26.

19. Develop an informal proof for Theorem T-27.

20. Develop an informal proof for Theorem T-28.

21. Develop an informal proof for Theorem T-29.

22. Develop an informal proof for Theorem T-30.

PART III

Some Topics in Euclidean and Non-Euclidean Plane Geometry

In this, the final part of this book, three topics will be presented. Chapter 9 contains an examination of some characteristics of Euclidean triangles and circles, while Chapter 10 contains an examination of the Euclidean concepts of cross ratios and harmonic sets of points and lines. Finally, Chapter 11 contains an examination of some topics from hyperbolic geometry, a truly non-Euclidean type of geometry which resulted from numerous attempts to prove that Euclid's parallel axiom was actually a theorem.

Since Part II of this book has as its basis David Hilbert's works, which were essentially a refinement of Euclidean geometry, much of the material in this Part III can be based on the material in Part II. In fact, unless new definitions and axioms are provided, the reader may assume that any terminology used in this part is the same as that in Part II. Additionally, it should be observed that the proofs of theorems presented in Part III can be developed, to a great extent, on the basis of materials from Part II and the standard Euclidean geometry materials presented in the Appendices.

Specifically, it is only in Chapter 11 that one axiom and a few of the theorems from Part II cannot be employed. This results from the fact that hyperbolic geometry permits the existence of more than one parallel for a line from an external point. Since this is contrary to the statement of Axiom 18 of Part II, it and all the theorems based on it must be disallowed in developing proofs for the theorems in Chapter 11.

Finally, the Supplementary Topics in this Part III are very abbreviated in form. The study of Transformation Geometry was concluded with the material at the end of Chapter 8. Also, the last set of Problems was presented at the end of Chapter 8. Only some relevant exercises in Constructions appear in this part. Since all of these pertain to Euclidean geometry, they are presented at the end of Chapter 9 and Chapter 10.

Chapter 9: Some Euclidean Geometry of Triangles and Circles

9.1 Introduction

One area of Euclidean geometry that has been of particular interest to modern geometers involves a collection of theorems pertaining to concurrent lines and another collection of theorems pertaining to collinear points. A study of these theorems will result in some observations concerning the Euclidean geometry of triangles and circles. While much of this information is based on ancient theorems, part of it involves observations by later mathematicians.

9.2 Some Theorems Concerning Concurrences in Triangles

Among the theorems from ancient Euclidean geometry are several which pertain to concurrences among various lines associated with a triangle. Four of these theorems will be considered.

Definition 1: A collection of three or more lines which intersect at the same point are called **concurrent lines**.

Language Rule 1: The common point of intersection for a collection of concurrent lines will be called the **point of concurrency** for the lines.

Theorem 1: The perpendicular bisectors of the sides of a triangle are concurrent.

Proof: Suppose the three lines k, m, n are the perpendicular bisectors, respectively, of the sides (A,B), (B,C), and (A,C) of $\triangle ABC$. (See **Figure 9.1**.)

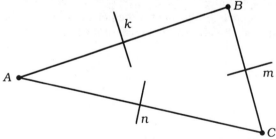

Figure 9.1

Using Part II material (Theorem 94), it can be shown that k and m must intersect at a point G, since k and m are perpendicular, respectively, to the lines AB and BC, which intersect at B.

Since G lies on k, then, from Part II material (Theorem 98), G is equidistant from each of A and B. Similarly, G is equidistant from each of B and C. Since $(A,G) \cong (B,G)$ and $(B,G) \cong (C,G)$, then, by Part II material (Axiom 16), $(A,G) \cong (C,G)$.

Hence, G is equidistant from each of A and C. Thus, by Part II material (Theorem 98), G lies on n. Therefore, k, m, n are concurrent. ∎

Definition 2: The set of all points which are located at a given distance from a given point is called a **circle**. The given point is called the **center** of the circle, and the given distance is called the **radius** of the circle. Any segment having the center for one endpoint and a point of the circle for its second endpoint is called a **radius** of the circle, also. Any segment which has for its endpoints two points of the circle is called a **chord** of the circle. A chord which contains the center of a circle is called a **diameter** of the circle.

Definition 3: A line which contains exactly two points of a circle is called a **secant line** of the circle. (See **Figure 9.2**.)

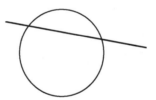

Figure 9.2

Definition 4: A line which contains exactly one point of a circle and is perpendicular to the line containing the radius having this point for an endpoint is called a **tangent line** of the circle. (See **Figure 9.3**.)

Figure 9.3

Language Rule 2: A tangent line of a circle can be described as being tangent to the circle. Also, the circle can be described as being **tangent to the line**.

In developing the proof for Theorem 1, the point of concurrency for the three perpendicular bisectors of the triangle's sides was shown to be equidistant from the vertices of the triangle. Hence, it can be observed, this point of concurrency can serve as the center of a circle which contains the triangle's three vertices. (See **Figure 9.4**.)

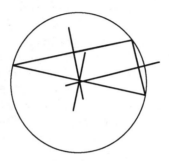

Figure 9.4

Definition 5: The point of concurrency of the perpendicular bisectors of the three sides of a triangle is called the **circumcenter** of the triangle.

Definition 6: The circle which contains the three vertices of a triangle and has for its center the circumcenter of the triangle is called the **circumcircle**, or **circumscribed circle**, of the triangle.

Definition 7: A line which contains a vertex of a triangle and is perpendicular to the sideline opposite the angle is called an **altitude line**. The segment which has for its endpoints the vertex and the point of intersection of the altitude line and the opposite sideline is called an **altitude**. (See **Figure 9.5**.)

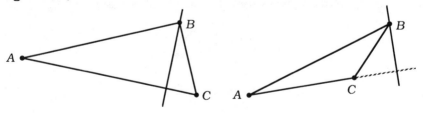

Figure 9.5

Theorem 2: The altitude lines of a triangle are concurrent.

Proof: Suppose k, m, n are the altitude lines of $\triangle ABC$, with $A \in k$, $B \in m$, and $C \in n$. (See **Figure 9.6**.)

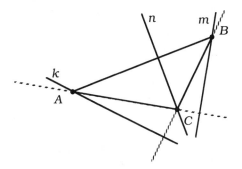

Figure 9.6

Using Part II material (Axiom 18 and Axiom 12), it can be shown that there exists $\triangle DEF$ such that lines DE and BC are parallel, lines DB and AC are parallel, A is the midpoint of (D,E), and F is the point of intersection of lines EC and DB. (See **Figure 9.7**.) Thus, lines EC and AB are transversals for parallel lines DE and BC and for parallel lines DB and AC. Also, $(E,A) \cong (A,D)$.

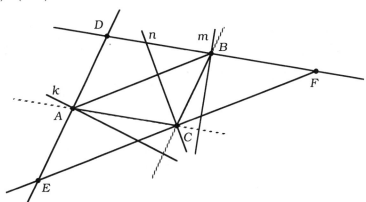

Figure 9.7

By Part II material (Theorem 79 and Theorem 81), $\angle DAB \cong \angle CBA$, $\angle DBA \cong \angle CAB$, $\angle AEC \cong \angle BCF$, and $\angle ACE \cong \angle BFC$.

Hence, in $\triangle ADB$ and $\triangle BCA$, $\angle DAB \cong \angle CBA$, $(A,B) \cong (B,A)$, and $\angle DBA \cong \angle CAB$. Thus, by Part II material (Theorem 55), $\triangle ADB \cong \triangle BCA$. Hence, $(D,A) \cong (C,B)$ and $(D,B) \cong (C,A)$.

Since $(E,A) \cong (A,D)$ and $(D,A) \cong (C,B)$, then, by Part II material (Axiom

13), $(E,A) \equiv (C,B)$.

In $\triangle EAC$ and $\triangle CBF$, $(E,A) \equiv (C,B)$, $\angle AEC \equiv \angle BCF$, and $\angle ACE \equiv \angle BFC$. Thus, by Part II material (Theorem 87), $\triangle EAC \equiv \triangle CBF$. Hence, $(A,C) \equiv (B,F)$ and $(E,C) \equiv (C,F)$.

Since $(D,B) \equiv (C,A)$ and $(A,C) \equiv (B,F)$, then, by Part II material (Axiom 13), $(D,B) \equiv (B,F)$.

Thus, B is the midpoint of (D,F) and C is the midpoint of (E,F), since $(D,B) \equiv (B,F)$ and $(E,C) \equiv (C,F)$.

By Part II material (Theorem 81 and Theorem 69), it can be shown that the altitude lines k, m, n are perpendicular, respectively, to lines ED, DF, and FE. Hence, k, m, n are, in fact, the perpendicular bisectors of the sides of $\triangle DEF$.

By Theorem 1, lines k, m, n are concurrent. ∎

Definition 8: The point of concurrency of the altitude lines of a triangle is called the **orthocenter** of the triangle.

Theorem 3: The bisectors of the angles of a triangle are concurrent.

Proof: Suppose $\overrightarrow{AD}$, $\overrightarrow{BE}$, and $\overrightarrow{CF}$ are the bisectors, respectively, of $\angle A$, $\angle B$, and $\angle C$ of $\triangle ABC$. (See **Figure 9.8**.) Thus, $D \in int(\angle BAC)$ and $E \in int(\angle ABC)$.

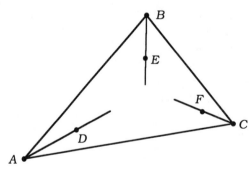

Figure 9.8

By Part II material (Theorem 41), $\overrightarrow{AD}$ intersects (B,C) at a point G. Since $G \in (B,C)$, then $G \in \overrightarrow{BC}$, $\overrightarrow{BG} = \overrightarrow{BC}$, and $\angle ABG \equiv \angle ABC$. Thus,

$E \in int(\angle ABG)$. Similarly, $\overrightarrow{BE}$ intersects (A,G) of $\triangle ABG$ at a point H. Therefore, $\overrightarrow{AD}$ and $\overrightarrow{BE}$ intersect at H.

Since H lies on the bisector of $\angle A$, then, from Appendices material, H is equidistant from the angle's sidelines. Similarly, H is equidistant from the sidelines of $\angle B$. By Part II material (Axiom 13), H is also equidistant from the sidelines of $\angle C$.

Thus, H is on the bisector of $\angle C$. Therefore, $\overrightarrow{AD}$, $\overrightarrow{BE}$, and $\overrightarrow{CF}$ are concurrent. ∎

In developing the proof for Theorem 3, the point of concurrency for the three bisectors of the triangle's angles was shown to be equidistant from the sidelines of the triangle. Using Part II material, it can be shown that this point lies in the interior of the triangle. Thus, the point is equidistant from the sides of the triangle.

Relative to the terminology of the Appendices material, the distance between a point and a line is described with the aid of a second line which contains the point and is perpendicular to the first line. That is, the distance is based on the segment having for its endpoints the original point and the point of intersection of the original line and the perpendicular line containing the original point. Thus, the point of concurrency of the angle bisectors of a triangle can serve as the center of a circle which contains one point from each side of the triangle. (See **Figure 9.9**.)

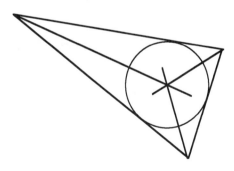

Figure 9.9

Definition 9: The point of concurrency of the bisector of the three angles of a triangle is called the **incenter** of the triangle.

Definition 10: The circle which has for its center the incenter of a triangle and is tangent to each side of the triangle is called the **incircle**, or **inscribed circle**, of the triangle.

Definition 11: A line which contains a vertex of a triangle and the midpoint of the opposite side of the triangle is called a **median line**. The segment which has for its endpoints the vertex and the midpoint of the opposite side is called a **median**. (See **Figure 9.10**.)

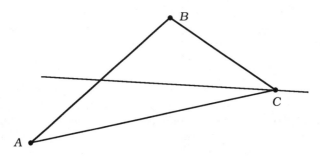

Figure 9.10

Theorem 4: The medians of a triangle are concurrent.

Definition 12: The point of concurrency of the medians of a triangle is called the **centroid** of the triangle.

Exercises 9.2

1. Under what conditions will the circumcenter of a triangle be in the interior of the triangle? on the triangle? in the exterior of the triangle? Explain.

2. Under what conditions will the orthocenter of a triangle be in the interior of the triangle? on the triangle? in the exterior of the triangle? Explain.

3. Under what conditions will the incenter of a triangle be in the interior of the triangle? on the triangle? in the exterior of the triangle? Explain.

4. Under what conditions will the centroid of a triangle be in the interior of the triangle? on the triangle? in the exterior of the triangle? Explain.

5. For a triangle, will the circumcenter, orthocenter, incenter, and centroid always be distinct points? Explain.

6. Develop a proof for Theorem 4. (Hint: Use Theorems 33, 38, 40, 41, and 55 of Part II along with Appendices material pertaining to similar triangles and parallelograms.)

7. Develop a proof for this statement:
The point of concurrency of the medians of a triangle partitions each median into two segments, one of which is twice as long as the other.

9.3 The Theorems of Ceva and Menelaus

Two well-known theorems concerning concurrence and collinearity for triangles are those of Ceva and Menelaus. About 1678, the Italian mathematician Ceva discovered his theorem pertaining to the concurrence of some lines associated with triangles. About the same time, he rediscovered the related theorem of the Greek mathematician Menelaus of Alexandria. It was about 100 B.C. that Menelaus presented his theorem pertaining to the collinearity of some points associated with a triangle.

Language Rule 3: For points on a line, one direction may be considered as being positive while the other direction is considered as being negative.

Definition 13: If segment (A,B) lies on a directed line, then (A,B) is called a **directed segment**.

Axiom 1: To each directed segment (A,B) there corresponds a unique real number. If the direction from A to B is positive, the real number is positive; if the direction from A to B is negative, the real number is negative.

Definition 14: The unique real number corresponding to a directed segment is called its **measure** or **length**.

Notation Rule 1: The measure of a directed segment (A,B) will be denoted as $m(A,B)$.

Axiom 2: If A, B, C are three points on a directed line, then
(a) $m(A,B) = -m(B,A)$;
(b) $m(A,B) + m(B,C) = m(A,C)$ if and only if $A \bullet B \bullet C$.

Definition 15: If a point P divides a directed segment (A,B) into two additional segments, the **ratio of division** is the number $\dfrac{m(A,P)}{m(P,B)}$. The point P is called a **dividing point** of the segment.

It should be observed that the ratio of division for a directed segment will be positive when the dividing point is in the segment. (See **Figure 9.11**.) This is true because both $m(A,P)$ and $m(P,B)$ will be positive or both will be negative.

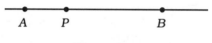

Figure 9.11

When the dividing point is not in the segment and is not an endpoint of the segment, the ratio of division will be negative. (See **Figure 9.12**.) This is true because one of $m(A,P)$ and $m(P,B)$ will be positive and the other will be negative.

Figure 9.12

Ratios of division of the sides of a triangle should be named by beginning at any vertex and proceeding around the triangle along directed segments in one fixed direction, either clockwise or counterclockwise. This movement around the triangle should proceed from the initially selected vertex to a dividing point, to the second vertex, to the next dividing point, to the third vertex, to the final dividing point, and return to the first vertex. This fixed-direction pattern is necessary to obtain appropriate signs for the lengths of the directed segments.

Theorem 5: (Ceva's Theorem) For a triangle, three lines, each containing a vertex and a point of the opposite sidelines of the triangle, are concurrent if and only if no two of the lines are parallel and the product of the ratios of division of the triangle's sides by the points is positive one (+1).

Proof: Suppose D, E, F are three points lying on the sidelines AB, BC, CA, respectively, of $\triangle ABC$ and that none of D, E, F is a vertex of $\triangle ABC$; without loss of generality, further suppose that $D \in (A,B)$, $E \in (B,C)$, and $F \in (C,A)$.

Claim: If lines AE, BF, CD are concurrent, then no two of them are parallel and $\dfrac{m(A,D)}{m(D,B)} \cdot \dfrac{m(B,E)}{m(E,C)} \cdot \dfrac{m(C,F)}{m(F,A)} = 1$.

Proof: Suppose AE, BF, CD are concurrent at a point G. (See **Figure 9.13**.)

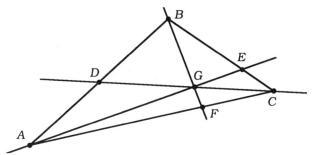

Figure 9.13

No two of AE, BF, CD are parallel since they all contain G.

Since $E \in (B,C)$, then by Part II material (Theorem 38), $E \in int(\angle BAC)$.

Since $F \in (C,A)$, then $F \in \overrightarrow{AC}$, $\overrightarrow{AF} = \overrightarrow{AC}$, $\angle BAF = \angle BAC$, and $E \in int(\angle BAF)$.

By Part II material (Theorem 41 and Theorem 2), $\overrightarrow{AE}$ intersects (B,F) at G. Thus, $G \in (B,F)$. Analogously, it can be shown that $G \in (A,E)$ and $G \in (C,D)$.

By Part II material (Axiom 18), there exists a line m which contains C and is parallel to line AB. Additionally, by Part II material (Theorem 82), each of AE and BF must intersect m; denote these points as J and H, respectively. (See **Figure 9.14**.)

Using Appendices material, $\triangle ADG$ is similar to $\triangle JCG$ and $\triangle HCG$ is similar to $\triangle BDG$. Thus, $\dfrac{m(A,D)}{m(J,C)} = \dfrac{m(D,G)}{m(C,G)}$ and $\dfrac{m(H,C)}{m(B,D)} = \dfrac{m(C,G)}{m(D,G)}$, respectively.

Therefore,

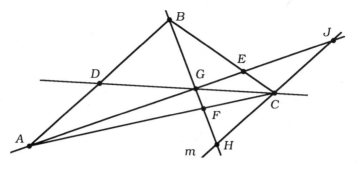

Figure 9.14

$$\frac{m(A,D)}{m(J,C)} \cdot \frac{m(H,C)}{m(B,D)} = \frac{m(D,G)}{m(C,G)} \cdot \frac{m(C,G)}{m(D,G)} = 1;$$

$$\frac{m(A,D)}{m(B,D)} \cdot \frac{m(H,C)}{m(J,C)} = 1; \quad \frac{m(A,D)}{-m(D,B)} \cdot \frac{-m(C,H)}{m(J,C)} = 1;$$

$$\text{and} \quad \frac{m(A,D)}{m(D,B)} = \frac{m(J,C)}{m(C,H)}.$$

Again using Appendices material, $\triangle BEA$ is similar to $\triangle CEJ$ and $\triangle CFH$ is similar to $\triangle AFB$.

Thus, $\dfrac{m(B,E)}{m(C,E)} = \dfrac{m(A,B)}{m(J,C)}$ and $\dfrac{m(C,F)}{m(A,F)} = \dfrac{m(H,C)}{m(B,A)}$, respectively.

Therefore, $\dfrac{m(A,D)}{m(D,B)} \cdot \dfrac{m(B,E)}{m(C,E)} \cdot \dfrac{m(C,F)}{m(A,F)} = \dfrac{m(J,C)}{m(C,H)} \cdot \dfrac{m(A,B)}{m(J,C)} \cdot \dfrac{m(H,C)}{m(B,A)};$

$$\frac{m(A,D)}{m(D,B)} \cdot \frac{m(B,E)}{-m(E,C)} \cdot \frac{m(C,F)}{-m(F,A)} = \frac{m(A,B)}{m(C,H)} \cdot \frac{-m(C,H)}{-m(A,B)}.$$

Thus, $\dfrac{m(A,D)}{m(D,B)} \cdot \dfrac{m(B,E)}{m(E,C)} \cdot \dfrac{m(C,F)}{m(F,A)} = 1.$ ♦

Claim: If $\dfrac{m(A,D)}{m(D,B)} \cdot \dfrac{m(B,E)}{m(E,C)} \cdot \dfrac{m(C,F)}{m(F,A)} = 1$ and no two of AE, BF, CD are parallel, then AE, BF, CD are concurrent.

Proof: Suppose $\dfrac{m(A,D)}{m(D,B)} \cdot \dfrac{m(B,E)}{m(E,C)} \cdot \dfrac{m(C,F)}{m(F,A)} = 1$ and no two of AE, BF, CD are parallel.

Since $E \in (B,C)$, then, by Part II material (Theorem 38), $E \in int(\angle BAC)$.

Since $D \in (A,B)$, then $D \in \overrightarrow{AB}$, $\overrightarrow{AD} = \overrightarrow{AB}$, $\angle DAC = \angle BAC$, and $E \in int(\angle DAC)$.

Thus, by Part II material (Theorem 41), $\overrightarrow{AE}$ intersects (D,C) at a point K. Hence, by Part II material (Theorem 38), $K \in int(\angle DAC)$. Also, $K \in (D,C)$ and, thus, $K \in \overrightarrow{CD}$.

By Part II material (Theorem 21 and Theorem 22), K is on the A-side of BC. Also, K is on the C-side of AD and K is on the D-side of AC. Thus, by Part II material (Theorem 21), K is on the C-side of AB and K is on the B-side of AC.

Hence, lines AE and CD intersect at a point K such that $K \in int(\triangle ABC)$. (See **Figure 9.15**.)

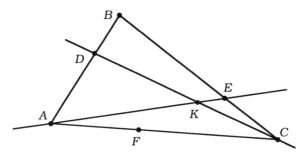

Figure 9.15

Since $K \in int(\triangle ABC)$, then $K \in int(\angle ABC)$. By Part II material (Theorem 41), $\overrightarrow{BK}$ intersects (A,C) at a point L. (See **Figure 9.16**.)

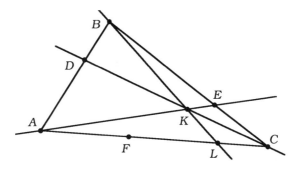

Figure 9.16

Therefore, the lines AE, BL, CD are concurrent at point K, and, by the preceding **Claim**, $\frac{m(A,D)}{m(D,B)} \cdot \frac{m(B,E)}{m(E,C)} \cdot \frac{m(C,L)}{m(L,A)} = 1$.

Since $\frac{m(A,D)}{m(D,B)} \cdot \frac{m(B,E)}{m(E,C)} \cdot \frac{m(C,F)}{m(F,A)} = 1$ and $\frac{m(A,D)}{m(D,B)} \cdot \frac{m(B,E)}{m(E,C)} \cdot \frac{m(C,L)}{m(L,A)} = 1$, then $\frac{m(A,D)}{m(D,B)} \cdot \frac{m(B,E)}{m(E,C)} \cdot \frac{m(C,F)}{m(F,A)} = \frac{m(A,D)}{m(D,B)} \cdot \frac{m(B,E)}{m(E,C)} \cdot \frac{m(C,L)}{m(L,A)}$.

Thus, $\frac{m(C,F)}{m(F,A)} = \frac{m(C,L)}{m(L,A)}$.

Since $F \in (A,C)$ and $L \in (A,C)$, then, by Part II material (Theorem 48), $F = L$. Thus $BL = BF$.

Hence, the lines AE, BF, CD are concurrent at K. ◆

Thus, three lines, each containing a vertex and a point of the opposite side of a triangle, are concurrent if and only if no two of the lines are parallel and the product of the ratios of division of the triangle's sides by the points is positive one. ■

Some observations need to be made concerning the statement and proof of the preceding theorem. First, the product mentioned in the theorem's statement would be meaningless if any of the three dividing points were a vertex. Also, Theorem 2, Theorem 20, and Theorem 41 of Part II insure that either all three sideline points must be on the triangle's sides or exactly one sideline point can be on a side. Finally, a similar proof could be developed for the situation in which exactly one sideline point is on a side.

Ceva's Theorem can be used in developing proofs of several theorems pertaining to concurrences of lines. One such theorem was developed early in the nineteenth century by the French mathematician J. D. Gergonne.

Theorem 6: (Gergonne's Theorem) For a triangle, the three lines determined by the vertices and the points of tangency of the triangle's incircle with the opposite sides are concurrent.

Proof: Suppose in $\triangle ABC$ the points of tangency of the triangle's incircle with the sides (A,B), (B,C), (C,A) are D, E, F, respectively. (See **Figure 9.17**.) From Part II material, the three lines AE, BF, CD exist.

Using Appendices material, $(A,D) \cong (F,A)$, $(B,E) \cong (D,B)$, and $(C,F) \cong (E,C)$, since they are pairs of tangent segments from the same point. Hence, $m(A,D) = m(F,A)$, $m(B,E) = m(D,B)$, and $m(C,F) = m(E,C)$.

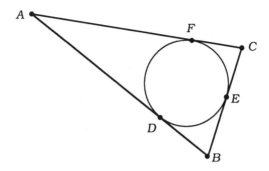

Figure 9.17

When D, E, F are considered as dividing points of segments (A,B), (B,C), (C,A), respectively,

$$\frac{m(A,D)}{m(D,B)} \cdot \frac{m(B,E)}{m(E,C)} \cdot \frac{m(C,F)}{m(F,A)} = \frac{m(A,D)}{m(F,A)} \cdot \frac{m(B,E)}{m(D,B)} \cdot \frac{m(C,F)}{m(E,C)} = 1.$$

Thus, by Theorem 5 (Ceva's Theorem), the lines AE, BF, and CD are concurrent. ∎

Theorem 7: (Menelaus' Theorem) For a triangle, three points, one lying on each sideline of the triangle, are collinear if and only if the product of the ratios of division of the triangle's sides by the points is negative one (–1).

Proof: Suppose D, E, F are three points lying on the sidelines AB, BC, CA, respectively, of $\triangle ABC$ and that none of D, E, F is a vertex of $\triangle ABC$; without loss of generality, further suppose that $D \in (A,B)$, $E \in (B,C)$, and $F \bullet A \bullet C$.

Claim: If points D, E, F are collinear, then

$$\frac{m(A,D)}{m(D,B)} \cdot \frac{m(B,E)}{m(E,C)} \cdot \frac{m(C,F)}{m(F,A)} = -1.$$

Proof: Suppose D, E, F are collinear on a line m. (See **Figure 9.18**.)

By Part II material (Theorem 90), there is a line which contains A and is perpendicular to m at a point G. Similarly, there is a line which contains B and is perpendicular to m at H. Also, there is a line which contains C and is perpendicular to m at J. (See **Figure 9.19**.)

Using Appendices material, $\triangle FCJ$ is similar to $\triangle FAG$, $\triangle BHE$ is similar to $\triangle CJE$, and $\triangle AGD$ is similar to $\triangle BHD$. Hence,

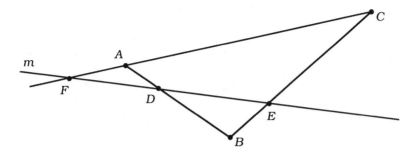

Figure 9.18

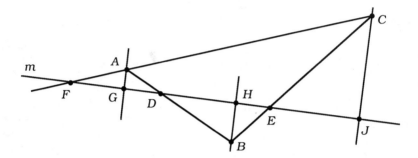

Figure 9.19

$$\frac{m(F,C)}{m(F,A)} = \frac{m(F,J)}{m(F,G)} = \frac{m(C,J)}{m(A,G)}, \frac{m(B,E)}{m(E,C)} = \frac{m(B,H)}{m(J,C)} = \frac{m(H,E)}{m(E,J)},$$

$$\text{and } \frac{m(A,D)}{m(D,B)} = \frac{m(A,G)}{m(H,B)} = \frac{m(G,D)}{m(D,H)}.$$

Thus, $\dfrac{m(A,D)}{m(D,B)} \cdot \dfrac{m(B,E)}{m(E,C)} \cdot \dfrac{m(C,F)}{m(F,A)} = \dfrac{m(A,G)}{m(H,B)} \cdot \dfrac{m(B,H)}{m(J,C)} \cdot \dfrac{m(C,J)}{m(A,G)}$

$$= \frac{m(A,G)}{m(H,B)} \cdot \frac{-m(H,B)}{-m(C,J)} \cdot -\frac{m(C,J)}{m(A,G)} = -1.$$

Therefore, $\dfrac{m(A,D)}{m(D,B)} \cdot \dfrac{m(B,E)}{m(E,C)} \cdot \dfrac{m(C,F)}{m(F,A)} = -1.$ ♦

Claim: If $\dfrac{m(A,D)}{m(D,B)} \cdot \dfrac{m(B,E)}{m(E,C)} \cdot \dfrac{m(C,F)}{m(F,A)} = -1$, then D, E, F are collinear.

Proof: Suppose $\dfrac{m(A,D)}{m(D,B)} \cdot \dfrac{m(B,E)}{m(E,C)} \cdot \dfrac{m(C,F)}{m(F,A)} = -1$.

From Part II material (Theorem 20, Theorem 21, Theorem 22, and Theorem 39), E and F determine a line m which intersects side (A,B) at a point G. Also, from Part II material, G cannot be a vertex of $\triangle ABC$. (See **Figure 9.20**.)

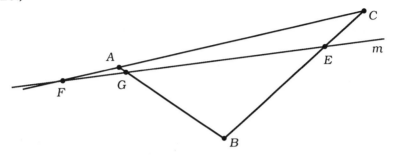

Figure 9.20

Thus, G, E, F are three collinear points lying on the sidelines AB, BC, CA, respectively, of $\triangle ABC$. From the preceding **Claim**,

$$\dfrac{m(A,G)}{m(G,B)} \cdot \dfrac{m(B,E)}{m(E,C)} \cdot \dfrac{m(C,F)}{m(F,A)} = -1.$$

Using the preceding equation and the equation in the supposition for this **Claim**, $\dfrac{m(A,D)}{m(D,B)} \cdot \dfrac{m(B,E)}{m(E,C)} \cdot \dfrac{m(C,F)}{m(F,A)} = \dfrac{m(A,G)}{m(G,B)} \cdot \dfrac{m(B,E)}{m(E,C)} \cdot \dfrac{m(C,F)}{m(F,A)}$.

Thus, $\dfrac{m(A,D)}{m(D,B)} = \dfrac{m(A,G)}{m(G,B)}$.

Hence, $\dfrac{m(A,D)}{m(D,B)} + \dfrac{m(D,B)}{m(D,B)} = \dfrac{m(A,G)}{m(G,B)} + \dfrac{m(G,B)}{m(G,B)}$; $\dfrac{m(A,B)}{m(D,B)} = \dfrac{m(A,B)}{m(G,B)}$.

Since $m(A,B) = m(A,B)$, then $m(D,B) = m(G,B)$.

Thus, (D,B) and (G,B) are collinear directed segments having the same length. From Part II material (Axiom 12), $D = G$.

Therefore, D, E, F are collinear. ♦

Thus, three points, one lying on each sideline of a triangle, are collinear if and only if the product of the ratios of division of the triangle's sides by the points is negative one. ∎

Some observations need to be made concerning the statement and proof of the preceding theorem. First, the product mentioned in the theorem's statement would be meaningless if any of the three dividing points were a vertex. Also, Axiom 11 and Theorem 18 of Part II insure that the line containing the dividing points must intersect exactly two of the triangle's sides, if it intersects any one of them. Finally, an analogous proof could be developed for the situation in which none of the dividing points are on the sides of the triangle.

The Theorem of Menelaus can be used in developing proofs of several theorems pertaining to collinearity of points. Two of the theorems will considered.

Theorem 8: If each of the lines which are tangent to a triangle's circumcircle at its vertices intersects the opposite sideline, the three points of intersection are collinear.

Proof: Suppose for $\triangle ABC$ the lines m, n, k are tangent to the triangle's circumcircle at A, B, C, respectively; further suppose that m, n, k intersect the triangle's sidelines BC, AC, AB at D, E, F, respectively. (See **Figure 9.21**.)

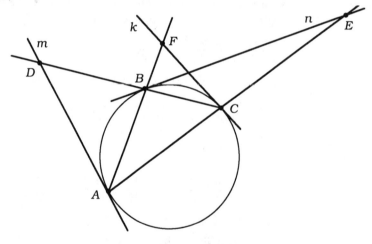

Figure 9.21

The points D, E, F can be considered as dividing points of segments (B,C), (C,A), (A,B), respectively. In addition, sidelines AB, AC, BC are secant lines of the circumcircle.

Claim: $\dfrac{m(A,E)}{m(E,C)} = -\dfrac{(m(A,B))^2}{(m(B,C))^2}.$

Proof: Using Appendices material pertaining to chords and tangent lines, $\angle EBA$ of $\triangle BEA$ is congruent to $\angle C$ of $\triangle BEC$. Also, $\angle E$ is in both triangles. Hence, by Appendices material, $\triangle BEA$ is similar to $\triangle CEB$.

Thus, $\dfrac{m(A,B)}{m(B,C)} = \dfrac{m(B,E)}{m(C,E)},$ and $\dfrac{(m(A,B))^2}{(m(B,C))^2} = \dfrac{(m(B,E))^2}{(m(C,E))^2}.$

Using Appendices material pertaining to tangent lines and secant lines, $m(B,E)$ is the mean proportional between $m(C,E)$ and $m(A,E)$.

Hence, $\dfrac{m(C,E)}{m(B,E)} = \dfrac{m(B,E)}{m(A,E)},$ and $(m(B,E))^2 = m(C,E) \cdot m(A,E).$

Therefore, $\dfrac{(m(A,B))^2}{(m(B,C))^2} = \dfrac{(m(B,E))^2}{(m(C,E))^2} = \dfrac{m(C,E) \cdot m(A,E)}{(m(C,E))^2},$

and $\dfrac{(m(A,B))^2}{(m(B,C))^2} = \dfrac{m(A,E)}{m(C,E)}.$

Hence, $\dfrac{(m(A,B))^2}{(m(B,C))^2} = -\dfrac{m(A,E)}{m(E,C)},$ and $\dfrac{m(A,E)}{m(E,C)} = -\dfrac{(m(A,B))^2}{(m(B,C))^2}.$ ♦

Claim: $\dfrac{m(C,D)}{m(D,B)} = -\dfrac{(m(C,A))^2}{(m(A,B))^2}.$

Proof: Analogous to the preceding **Claim.** ♦

Claim: $\dfrac{m(B,F)}{m(F,A)} = -\dfrac{(m(B,C))^2}{(m(C,A))^2}.$

Proof: Analogous to the preceding **Claim.** ♦

Therefore,

$$\frac{m(A,E)}{m(E,C)} \cdot \frac{m(C,D)}{m(D,B)} \cdot \frac{m(B,F)}{m(F,A)} = -\frac{(m(A,B))^2}{(m(B,C))^2} \cdot \frac{(m(C,A))^2}{(m(A,B))^2} \cdot \frac{(m(B,C))^2}{(m(C,A))^2} = -1.$$

Hence, $\dfrac{m(A,E)}{m(E,C)} \cdot \dfrac{m(C,D)}{m(D,B)} \cdot \dfrac{m(B,F)}{m(F,A)} = -1.$

By Theorem 7 (Menelaus' Theorem), the points D, E, F are collinear. ∎

Theorem 9: If the line containing the bisector of an exterior angle of a triangle intersects the sideline opposite its vertex, this point of intersection is collinear with the points of intersection of the bisectors of the remote interior angles at the other vertices and their opposite sidelines.

Exercises 9.3

1. Use Ceva's Theorem to develop a proof for this statement:
The median lines of a triangle are concurrent.

2. Use Ceva's Theorem to develop a proof for Theorem 3.

3. Use Ceva's Theorem to develop a proof for this statement:
If E is the midpoint of side (B,C) of $\triangle ABC$ and D and F are points on sides (A,B) and (C,A), respectively, such that lines CD and BF are concurrent with line AE, then line DF is parallel to (B,C).

4. Use Gergonne's Theorem, Ceva's Theorem, and Menelaus' Theorem to develop a proof for this statement:
If the incircle of $\triangle ABC$ is tangent to the sides (A,B), (B,C), (C,A) at the points X, Y, Z, respectively, and the line ZY intersects the sideline AB at K, then $\dfrac{m(A,X)}{m(X,B)} = \dfrac{m(A,K)}{m(B,K)}$.

5. Use Ceva's Theorem to develop a proof for this statement:
If D and E are points on sides (A,B) and (B,C), respectively, of $\triangle ABC$ such that (D,E) is parallel to (A,C) and lines AE and CD intersect at point K, then line BK is a median line for $\triangle ABC$.

6. Develop a proof for Theorem 9.

7. Use Menelaus' Theorem to develop a proof for this statement:
   If a line intersects the sides (A,B), (B,C), (C,D), (D,A) of a quadrilateral $ABCD$ in points E, F, G, H, respectively, then
   $$\frac{m(A,E)}{m(E,B)} \cdot \frac{m(B,F)}{m(F,C)} \cdot \frac{m(C,G)}{m(G,D)} \cdot \frac{m(D,H)}{m(H,A)} = 1.$$

8. Use Menelaus' Theorem to prove the first **Claim** in the proof of Ceva's Theorem.

9.4 The Simson Line, the Nine-Point Circle, and the Euler Line

A study of Euclidean geometry of triangles and circles would be somewhat incomplete without an examination of some observations made by the English mathematician Simson, the German mathematician Feuerbach, and the Swiss mathematician Euler.

Simson, who lived from 1687 to 1768, discovered the following theorem pertaining to the collinearity of some points associated with a triangle.

Theorem 10: (Simson's Theorem) For a triangle, the lines through any point of its circumcircle and perpendicular to its sidelines intersect the sidelines in collinear points.

Proof: Suppose S is a point on the circumscribed circle of $\triangle ABC$; further suppose that line m containing S and perpendicular to sideline AB intersects AB at D, line n containing S and perpendicular to sideline BC intersects BC at E, and line k containing S and perpendicular to sideline CA intersects CA at F.

There are two possible locations for S:
(i) S is a vertex of $\triangle ABC$;
(ii) S is not a vertex of $\triangle ABC$.

Claim: If S is a vertex of $\triangle ABC$, then D, E, F are collinear.

Proof: Left as an exercise. ♦

Claim: If S is not a vertex of $\triangle ABC$, then D, E, F are collinear.

Proof: There are two possible locations for D, E, F:
(i) One of D, E, F is a vertex of $\triangle ABC$;
(ii) None of D, E, F is a vertex of $\triangle ABC$.

Subclaim: If one of D, E, F is a vertex of △ABC, then D, E, F are collinear.

Proof: Left as an exercise. ♦♦

Subclaim: If none of D, E, F is a vertex of △ABC, then D, E, F are collinear.

Proof: Suppose S is not a vertex of △ABC and none of D, E, F is a vertex of △ABC. (See **Figure 9.22**.)

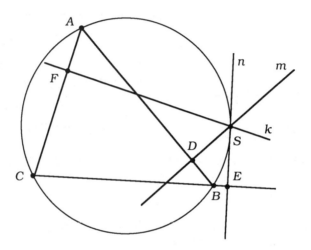

Figure 9.22

Since each of ∠SDB and ∠SEB is a right angle, then, by Part II material, they are supplementary opposite angles of quadrilateral SEBD. Also, ∠ESD and ∠EBD are supplementary opposite angles. Hence, by Appendices material, quadrilateral SEBD can be inscribed in a circle. (See **Figure 9.23**.)

Using Appendices material pertaining to inscribed angles, m∠SED = m∠SBD. Hence, ∠SED ≅ ∠SBD.

Analogously, in quadrilateral SECF, ∠SCF ≅ ∠SEF. (See **Figure 9.23**.)
Analogously, in quadrilateral SBCA, ∠SBA ≅ ∠SCA. (See **Figure 9.23**.)
Using Part II material (Axiom 16),

∠SED ≅ ∠SBD = ∠SBA ≅ ∠SCA = ∠SCF ≅ ∠SEF.

Hence, ∠SED ≅ ∠SEF. Using Part II material (Axiom 15), $\overrightarrow{ED} = \overrightarrow{EF}$. Thus, D, E, F are collinear. ♦♦

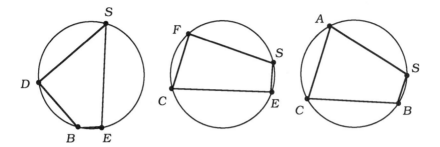

Figure 9.23

Therefore, if S is not a vertex of △ABC, then D, E, F are collinear. ♦

Thus, for a triangle, the lines through any point of its circumcircle and perpendicular to its sidelines intersect the sidelines in collinear points. ∎

Definition 16: A line containing the collinear points described in Simson's Theorem is called a **Simson line** for a triangle.

While there is not universal agreement among mathematical historians, it is believed by some that Feuerbach discovered the following theorem about 1822. This theorem describes a circle which contains some points associated with a triangle. It should be observed that, under some circumstances, the points considered in the theorem may not be distinct. This, however, does not diminish the significance of the theorem, whose proof will be presented for the situation in which all the points are distinct.

Theorem 11: (Nine-Point Circle Theorem) For a triangle, the midpoints of its sides, the points of intersection of its altitude lines and its sidelines, and the midpoints of the segments from the orthocenter to the vertices all lie on a circle.

Proof: Suppose for △ABC, D, E, F are the midpoints, respectively, of sides (A,B), (B,C), (C,A); G, H, J are the points of intersection, respectively, of the altitude line through A and sideline BC, the altitude line through B and sideline AC, the altitude line through C and sideline AB; L, M, N are the midpoints, respectively, of the segments (A,K), (B,K), (C,K), where K is the orthocenter. (See **Figure 9.24**.) Further suppose that A, B, C, D, E, F, G, H, J, K, L, M, N are distinct.

From Appendices material, the three noncollinear points D, E, F lie on a circle Z.

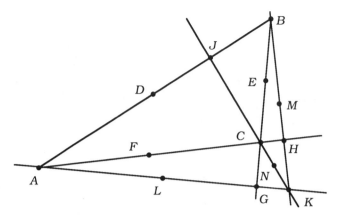

Figure 9.24

Claim: Points G, H, J lie on circle Z.

Proof: Since D and F are the midpoints, respectively, of sides (A,B) and (A,C) of △ABC, then, by Appendices material, line DF is parallel to sideline BC, which contains G. Hence, D, E, G, F are the vertices of a trapezoid.

By Appendices material, $m(D,E) = 0.5m(A,C)$, since D and E are the midpoints, respectively, of (A,B) and (B,C) of △ABC. Since F is the midpoint of the hypotenuse of right △ACG, then $m(F,G) = 0.5m(A,C)$. Hence, $m(D,E) = m(F,G)$, and $(D,E) \cong (F,G)$. Thus, D, E, G, F are the vertices of an isosceles trapezoid.

By Appendices material, the trapezoid DEGF can be inscribed in a circle, since it is a quadrilateral whose opposite angles are supplementary. Since D, E, F must lie together on a unique circle, G lies on circle Z.

Analogously, it can be shown that each of H and J lies on circle Z. ◆

Claim: Points L, M, N lie on circle Z.

Proof: Since F and L are the midpoints, respectively, of sides (A,C) and (A,K) of △ACK, then, by Appendices material, line FL is parallel to line CK. Similarly, line FE is parallel to line AB. Since CK is perpendicular to AB, then FL is perpendicular to FE, by Part II material (Theorem 93). Thus, △FEL is a right triangle.

Hence, by Appendices material, the two right triangles △FEL and △GEL can be inscribed in one circle having (E,L) as a diameter. Thus, this circle would contain the three points G, F, E, which already lie on the unique circle Z. Therefore, L lies on circle Z.

Analogously, it can be shown that each of M and N lies on circle Z. ♦

Therefore, all the points D, E, F, G, H, J, L, M, N lie on the unique circle Z. (See **Figure 9.25**.) ∎

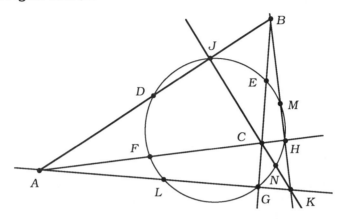

Figure 9.25

Definition 17: The circle containing the points described in the Nine-Point Circle theorem is called the **nine-point circle** for a triangle.

As the remaining material in this section will indicate, a triangle's nine-point circle has several interesting characteristics. Although some of the points associated with a triangle and its nine-point circle may not be distinct, the remaining proofs will be presented for situations in which the points are distinct.

Theorem 12: For a triangle, the radius of its nine-point circle is half the radius of its circumcircle.

Theorem 13: For a triangle, the center of its nine-point circle is the midpoint of the segment determined by its circumcenter and orthocenter.

Proof: Suppose for $\triangle ABC$, P is its circumcenter, K is its orthocenter, D is the midpoint of (A,B), E is the midpoint of (B,C), F is the midpoint of (A,C), and N is the midpoint of (K,C). (See **Figure 9.26**.)

By Theorem 12, each of D, E, and N is on the nine-point circle of $\triangle ABC$.

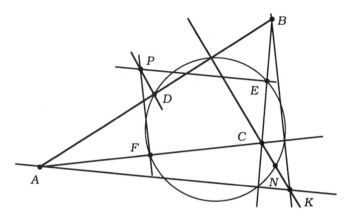

Figure 9.26

Claim: Quadrilateral PDKN is a parallelogram.

Proof: Since each of lines PD and NK is perpendicular to sideline AB, then, by Part II material (Theorem 78), PD and NK are parallel. Hence, (P,D) and (N,K) are parallel.

By Appendices material, $\triangle FDE$ is similar to $\triangle ACB$, with the similarity ratio being $1:2$. Also, PD, PE, PF are perpendicular, respectively, to sidelines FE, FD, DE of $\triangle FDE$. Thus, P is the orthocenter of $\triangle FDE$.

By Appendices material, $m(P,D) = 0.5m(C,K)$, since the similarity ratio for $\triangle FDE$ and $\triangle ACB$ is $1:2$. Since N is the midpoint of (K,C), then $m(P,D) = m(N,K)$. Thus, $(P,D) \cong (N,K)$.

Therefore, by Appendices material, quadrilateral PDKN is a parallelogram since one pair of opposite sides are parallel and congruent. ◆

Claim: (D,N) is a diameter of the nine-point circle.

Proof: For $\triangle ABC$, (D,E) is parallel to (A,C), by Appendices material. Analogously, for $\triangle BCK$, (E,N) is parallel to (B,K). Since line AC is perpendicular to line BK, then, by Part II material (Theorem 93), line DE is perpendicular to line EN.

Thus, $\angle DEN$ is a right angle, and $\triangle DEN$ is a right triangle inscribed in the nine-point circle. By Appendices material, the hypotenuse (D,N) of $\triangle DEN$ is a diameter of the nine-point circle. ◆

By Appendices material, the diagonals (P,K) and (D,N) of parallelogram PDKN bisect each other; denote this common midpoint as Y. Since (D,N) is a diameter of the nine-point circle, then Y is the center of the nine-point circle.

Therefore, the center of a triangle's nine-point circle is the midpoint of the segment determined by its circumcenter and orthocenter. ∎

Theorem 14: For a triangle having point P for its circumcenter, point K for its orthocenter, and point T for its centroid, T divides the directed segment (P,K) such that $\dfrac{m(P,T)}{m(P,K)} = \dfrac{1}{3}$.

Proof: Suppose for $\triangle ABC$, P is its circumcenter, K is its orthocenter, T is its centroid, D is the midpoint of (A,B), E is the midpoint of (B,C), and F is the midpoint of (C,A). (See **Figure 9.27**.)

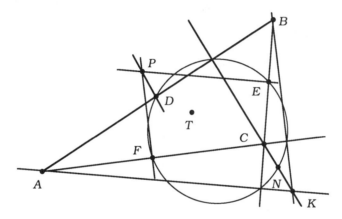

Figure 9.27

By Appendices material, $\triangle FDE$ is similar to $\triangle ACB$, with the similarity ratio being $1:2$. Also, (F,E) is parallel to (A,B), (F,D) is parallel to (C,B), and (D,E) is parallel to (A,C). By Part II material (Theorem 93), PD, PE, PF are perpendicular, respectively, to sidelines FE, FD, DE of $\triangle FDE$. Thus, P is the orthocenter for $\triangle FDE$. Hence, $m(P,D) = 0.5m(C,K)$.

Since each of lines PD and CK is perpendicular to sideline AB, then, by Part II material (Theorem 78), PD and CK are parallel lines. Hence, P, D, K, C are the vertices of a convex quadrilateral. By Appendices material, (P,K) and (D,C) intersect at a point X. Therefore, by Appendices material, $\triangle PDX$ is similar to $\triangle KCX$, with the similarity ratio being $1:2$, since

$m(P,D) = 0.5m(C,K)$.

Therefore, $m(D,X) = 0.5m(X,C)$ and $m(P,X) = 0.5m(X,K)$. Hence, $\dfrac{m(D,X)}{m(D,C)} = \dfrac{1}{3}$ and $\dfrac{m(P,X)}{m(P,K)} = \dfrac{1}{3}$. Since X lies on median (D,C) of $\triangle ABC$, and is located one-third of the distance from the midpoint D of side (A,B) to the opposite vertex C, then, by Appendices material, X is the centroid of $\triangle ABC$. Hence, $X = T$.

Thus, T divides the directed segment (P,K) such that $\dfrac{m(P,T)}{m(P,K)} = \dfrac{1}{3}$. ∎

The preceding theorem was discovered by Euler in 1765. Thus, the following definition.

Definition 18: For a triangle, if a unique line contains its orthocenter, circumcenter, centroid, and the center of its nine-point circle, it is called the **Euler line** for the triangle.

Exercises 9.4

1. Complete the proof for Theorem 10.

2. Develop a proof for Theorem 12.

3. For an equilateral triangle,
   (a) how many points of intersection will it have with its nine-point circle? Explain.
   (b) how is its nine-point circle related to its incircle? Explain.
   (c) does it have an Euler line? Explain.

4. For an isosceles triangle,
   (a) how many points of intersection will it have with its nine-point circle? Explain.
   (b) how is its nine-point circle related to its incircle? Explain.
   (c) does it have an Euler line? Explain.

5. For a right triangle,
   (a) how many points of intersection will it have with its nine-point circle? Explain.
   (b) how is its nine-point circle related to its incircle? Explain.

(c) does it have an Euler line? Explain.

6. In **Figure 9.28**, △ABC is shown with its circumcircle and nine-point circle. For △ABC, its circumcenter is at D, the center of its nine-point circle is at E, and its orthocenter is at F. In addition, points G and H are on the nine-point circle and circumcircle, respectively. Use this information to develop a proof for the following statement:

Point G is the midpoint of segment (F,H).

Note: This problem could have been formally stated in the following manner:

Develop a proof for this statement:

For a triangle, its nine-point circle bisects each segment determined by its orthocenter and a point on its circumcircle.

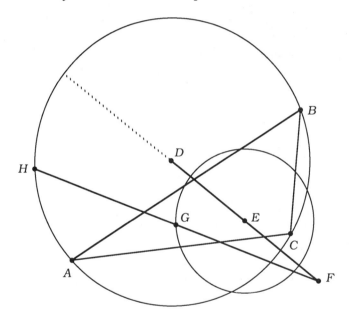

Figure 9.28

****

SUPPLEMENTARY TOPICS

CONSTRUCTIONS

38. Construct a triangle similar to the one given below. Use constructions to determine the triangle's circumcenter, incenter, orthocenter, centroid, nine-point circle's center, and Euler line.

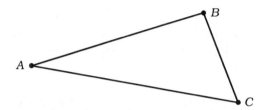

39. Construct a diagram similar to the one below. Use constructions to determine △ABC's Simson Line based on point D.

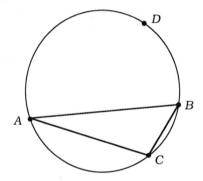

40. In the following diagram, points A and B are vertices of △ABC and point D is the center of the nine-point circle for △ABC. Construct a diagram similar to this one and use constructions to determine △ABC.

• A

D •

• B

41. In the following diagram, point A is a vertex of △ABC, point D is the orthocenter of △ABC, and point E is the circumcenter of △ABC. Construct a diagram similar to this one and use constructions to determine △ABC.

• D

• A

• E

42. In the following diagram, lines m and n are parallel. Construct a diagram similar to this one and use only a straightedge to bisect (A,B).

n

Chapter 10: Cross Ratio and Harmonic Sets

10.1 Introduction

This chapter presents an examination of relationships existing among sets of four collinear points and relationships existing among sets of four lines which intersect at the same point. Various materials from Chapter 9, Part II, and standard Euclidean geometry will be used in developing the concepts of this chapter.

10.2 Cross Ratios of Points and Lines

The relationships existing among a set of four collinear points will be developed initially. This material will be used to describe relationships existing among sets of four lines which intersect at a point.

Definition 1: If A, B, C, D are four collinear points, the **cross ratio** of these points is the number

$$\lambda = \frac{\text{ratio into which } C \text{ divides } (A,B)}{\text{ratio into which } D \text{ divides } (A,B)} = \frac{m(A,C)/m(C,B)}{m(A,D)/m(D,B)}.$$

It should be observed that the cross ratio of four collinear points may be either positive or negative, since it is defined in terms of the measure of directed segments.

Notation Rule 1: The cross ratio of four collinear points A, B, C, D will sometimes be denoted by (AB,CD). In the symbol (AB,CD), points C and D are dividing points of segment (A,B).

Although there are twenty-four permutations of four collinear points and, hence, twenty-four possible cross ratios, there will not be twenty-four distinct cross ratios. The following collection of theorems will present observations concerning the relationships among the various permutations. These theorems will show that order is an essential part of the definition of cross ratio.

Theorem 1: If A, B, C, D are four collinear points, then
$$(AB,CD) = (CD,AB) = (BA,DC) = (DC,BA).$$

Proof: Suppose A, B, C, D are four collinear points.
By Definition 1 and the material on directed segments in Chapter 9,

$$(AB,CD) = \frac{m(A,C)/m(C,B)}{m(A,D)/m(D,B)} = \frac{m(A,C) \cdot m(D,B)}{m(C,B) \cdot m(A,D)} = \frac{-m(A,C) \cdot -m(D,B)}{m(C,B) \cdot m(A,D)}$$

$$= \frac{m(C,A) \cdot m(B,D)}{m(A,D) \cdot m(C,B)} = \frac{m(C,A)/m(A,D)}{m(C,B)/m(B,D)} = (CD, AB).$$

Analogously, it can also be shown that $(CD,AB) = (BA,DC)$ and $(BA,DC) = (DC,BA)$.

Hence, $(AB,CD) = (CD,AB) = (BA,DC) = (DC,BA)$. ∎

Theorem 2: If A, B, C, D are four collinear points and $(AB,CD) = \lambda$, then $(BA,CD) = \dfrac{1}{\lambda}$.

Theorem 3: If A, B, C, D are four collinear points and $(AB,CD) = \lambda$, then $(AC,BD) = 1 - \lambda$.

Proof: Suppose A, B, C, D are four collinear points and $(AB,CD) = \lambda$.
By Definition 1 and the material on directed segments in Chapter 9,

$$(AC,BD) = \frac{m(A,B)/m(B,C)}{m(A,D)/m(D,C)} = \frac{m(A,B) \cdot m(D,C)}{m(B,C) \cdot m(A,D)}$$

$$= \frac{m(A,C) + m(C,B)}{m(B,C)} \cdot \frac{m(D,B) + m(B,C)}{m(A,D)}$$

$$= \left(\frac{m(A,C)}{m(B,C)} + \frac{m(C,B)}{m(B,C)}\right)\left(\frac{m(D,B)}{m(A,D)} + \frac{m(B,C)}{m(A,D)}\right)$$

$$= \frac{m(A,C)}{m(B,C)} \cdot \frac{m(D,B)}{m(A,D)} + \frac{m(A,C)}{m(B,C)} \cdot \frac{m(B,C)}{m(A,D)} + \frac{m(C,B)}{m(B,C)} \cdot \frac{m(D,B)}{m(A,D)} + \frac{m(C,B)}{m(B,C)} \cdot \frac{m(B,C)}{m(A,D)}$$

$$= \frac{m(A,C)}{m(B,C)} \cdot \frac{m(D,B)}{m(A,D)} + \frac{m(B,C)}{m(B,C)}\left(\frac{m(A,C)}{m(A,D)} + \frac{-m(D,B)}{m(A,D)} + \frac{m(C,B)}{m(A,D)}\right)$$

$$= \frac{m(A,C)}{-m(C,B)} \cdot \frac{m(D,B)}{m(A,D)} + \frac{m(A,C)}{m(A,D)} + \frac{m(C,B)}{m(A,D)} + \frac{m(B,D)}{m(A,D)}$$

$$= -\frac{m(A,C)/m(C,B)}{m(A,D)/m(D,B)} + \frac{m(A,C)}{m(A,D)} + \frac{m(C,D)}{m(A,D)} = -\lambda + \frac{m(A,D)}{m(A,D)} = -\lambda + 1 = 1 - \lambda.$$

Hence, $(AC,BD) = 1 - \lambda$. ∎

Theorem 4: If A, B, C, D are four collinear points and $(AB,CD) = \lambda$, then $(BC,AD) = \dfrac{\lambda - 1}{\lambda}$.

Proof: Suppose A, B, C, D are four collinear points and $(AB,CD) = \lambda$. By Theorem 3, $(BA,CD) = 1 - (BC,AD)$. Thus, $(BC,AD) = 1 - (BA,CD)$. By Theorem 2, $(BA,CD) = \dfrac{1}{\lambda}$, since $(AB,CD) = \lambda$.

Thus, $(BC,AD) = 1 - \dfrac{1}{\lambda} = \dfrac{\lambda - 1}{\lambda}$. ∎

Theorem 5: If A, B, C, D are four collinear points and $(AB,CD) = \lambda$, then $(CA,BD) = \dfrac{1}{1 - \lambda}$.

Theorem 6: If A, B, C, D are four collinear points and $(AB,CD) = \lambda$, then $(CB,AD) = \dfrac{\lambda}{\lambda - 1}$.

The information contained in the preceding six theorems can be used to prove the next theorem. This theorem will provide the final observations concerning the set of cross ratios associated with four collinear points.

Theorem 7: If A, B, C, D are four collinear points and $(AB,CD) = \lambda$, then

$$(BA,DC) = (CD,AB) = (DC,BA) = \lambda,$$

$$(AB,DC) = (BA,CD) = (CD,BA) = (DC,AB) = \dfrac{1}{\lambda},$$

$$(AC,BD) = (BD,AC) = (CA,DB) = (DB,CA) = 1 - \lambda,$$

$$(AD,BC) = (BC,AD) = (CB,DA) = (DA,CB) = \dfrac{\lambda - 1}{\lambda},$$

$$(AC,DB) = (BD,CA) = (CA,BD) = (DB,AC) = \dfrac{1}{1 - \lambda},$$

and $(AD,CB) = (BC,DA) = (CB,AD) = (DA,BC) = \dfrac{\lambda}{\lambda - 1}$.

Theorem 8: For four points A, B, C, D on a line n and a point P not on

n, $(AB,CD) = \lambda$ if and only if $\dfrac{m(B,F)}{m(B,E)} = \lambda$, where E and F are the points where a line k parallel to line AP and containing B intersects lines PC and PD, respectively.

Proof: Suppose A, B, C, D are four points on a line n and P is a point not on n; without loss of generality, suppose $A \bullet C \bullet B \bullet D$. (See **Figure 10.1**.) Further suppose a line k parallel to line AP, containing B, and intersecting lines PC and PD in points E and F, respectively.

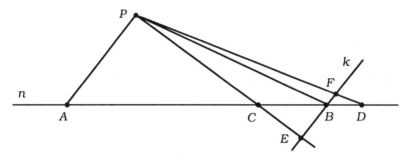

Figure 10.1

Claim: If $(AB,CD) = \lambda$, then $\dfrac{m(B,F)}{m(B,E)} = \lambda$.

Proof: Suppose $(AB,CD) = \lambda$.

In similar triangles $\triangle APD$ and $\triangle BFD$, $\dfrac{m(A,D)}{m(B,D)} = \dfrac{m(A,P)}{m(B,F)}$. Thus, $\dfrac{m(B,D)}{m(A,D)} = \dfrac{m(B,F)}{m(A,P)}$.

In similar triangles $\triangle APC$ and $\triangle BEC$, $\dfrac{m(A,C)}{m(C,B)} = \dfrac{m(A,P)}{m(E,B)}$.

Hence, $\dfrac{m(A,C)}{m(C,B)} \cdot \dfrac{m(B,D)}{m(A,D)} = \dfrac{m(A,P)}{m(E,B)} \cdot \dfrac{m(B,F)}{m(A,P)}$;

$\dfrac{m(A,C)}{m(C,B)} \cdot \dfrac{-m(D,B)}{m(A,D)} = \dfrac{m(B,F)}{-m(B,E)}$; $\dfrac{m(A,C)/m(C,B)}{m(A,D)/m(D,B)} = \dfrac{m(B,F)}{m(B,E)}$.

Since $\dfrac{m(A,C)/m(C,B)}{m(A,D)/m(D,B)} = (AB,CD) = \lambda$, then $\dfrac{m(B,F)}{m(B,E)} = \lambda$. ♦

Claim: If $\dfrac{m(B,F)}{m(B,E)} = \lambda$, then $(AB,CD) = \lambda$.

Proof: Left as an exercise. ◆

Thus, for four points A, B, C, D on a line n and a point P not on n, $(AB,CD) = \lambda$ if and only if $\dfrac{m(B,F)}{m(B,E)} = \lambda$, where E and F are the points where a line k parallel to line AP and containing B intersects lines PC and PD, respectively. ∎

Definition 2: If m and n are two lines and P is a point on neither line, then the **central projection** of a point A of m onto n is the point A' at which line PA intersects n. The point P is called the **center of projection**, and A' is called the **image** of A. (See **Figure 10.2**.)

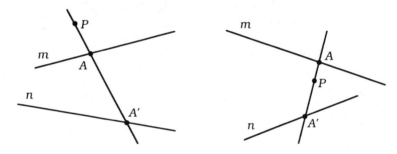

Figure 10.2

Theorem 9: If four collinear points A, B, C, D having cross ratio $(AB,CD) = \lambda$ are centrally projected onto a second line, then $(A'B',C'D') = \lambda$, where A', B', C', D' are the images of A, B, C, D, respectively.

Proof: Suppose A, B, C, D are four points on a line n, $(AB,CD) = \lambda$, k is a second line, and P is a point on neither n nor k; without loss of generality, suppose $A \bullet C \bullet B \bullet D$. (See **Figure 10.3**.) Further suppose points A', B', C', D' on k are the images of A, B, C, D, respectively, resulting from a projection centered at P.

Suppose s is a line through B parallel to line PA and intersecting lines PC and PD at points E and F, respectively. (See **Figure 10.4**.) Also, suppose t is a line through B' parallel to line PA and intersecting lines PC' and PD' at points E' and F', respectively.

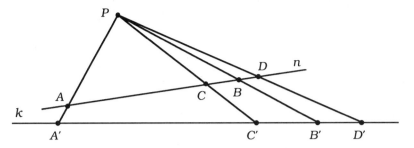

Figure 10.3

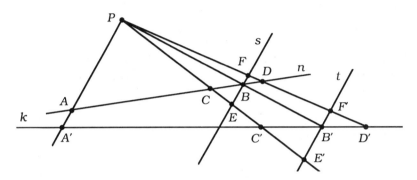

Figure 10.4

By Theorem 8, $\dfrac{m(B,F)}{m(B,E)} = \lambda$.

In similar triangles $\triangle FBP$ and $\triangle F'B'P$, $\dfrac{m(B,F)}{m(B',F')} = \dfrac{m(P,B)}{m(P,B')}$.

In similar triangles $\triangle EBP$ and $\triangle E'B'P$, $\dfrac{m(B,E)}{m(B',E')} = \dfrac{m(P,B)}{m(P,B')}$.

Thus, $\dfrac{m(B,F)}{m(B',F')} = \dfrac{m(B,E)}{m(B',E')}$. Hence, $\dfrac{m(B,F)}{m(B,E)} = \dfrac{m(B',F')}{m(B',E')}$.

Therefore, $\dfrac{m(B',F')}{m(B',E')} = \lambda$. By Theorem 8, $(A'B',C'D') = \lambda$. ∎

Definition 3: A **range of points** is a collection of two or more collinear points. The **base of a range** is the line on which the points lie.

Definition 4: A set of concurrent lines is called a **pencil of lines**, and the point through which the lines pass is called the **vertex of the pencil**.

Definition 5: The **cross ratio of a pencil** of four lines is the cross ratio of the range of points of intersection of these lines with any transversal which does not contain the vertex of the pencil.

Notation Rule 2: The cross ratio of a pencil of four lines having a point P for its vertex will sometimes be denoted as $P(AB,CD)$, where (AB,CD) is the cross ratio of the range of points of intersection of these lines with a transversal.

Theorem 9 and the preceding definitions can be used to produce proofs for the following pair of theorems.

Theorem 10: If the four lines of a pencil having P for its vertex are intersected by two transversals in the points A, B, C, D and A', B', C', D', respectively, then $P(AB,CD) = P(A'B',C'D')$. (See **Figure 10.5**.)

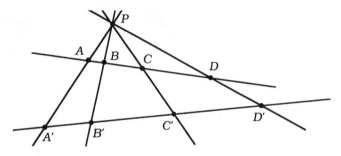

Figure 10.5

Theorem 11: If the four lines of a pencil having a point P for its vertex and the four lines of a pencil having a second point Q for its vertex are intersected by a transversal in the same four points A, B, C, D, then $P(AB,CD) = Q(AB,CD)$. (See **Figure 10.6**.)

Exercises 10.2

1. Develop a proof for Theorem 2.

2. Develop a proof for Theorem 5.

3. Develop a proof for Theorem 6.

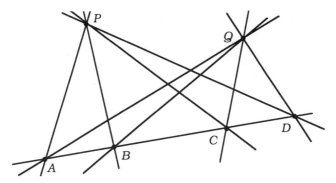

Figure 10.6

4. Develop a proof for Theorem 7.

5. Complete the proof for Theorem 8.

6. Develop a proof for Theorem 10.

7. Develop a proof for Theorem 11.

8. Develop a proof for this statement:
   If A, B, C, D are four collinear points, $(AB,CD) = \lambda$, and $(AB,CD) = (BA,CD) = (AB,DC)$, then $\lambda = -1$.

9. Develop a proof for this statement:
   If the four lines of a pencil having a point P for its vertex are intersected by a transversal at the points A, B, C, D and a line through B parallel to line PC intersects lines PA and PD at the points A' and D', respectively, then $P(BC,AD) = \dfrac{m(A',B)}{m(D',B)}$.

10. Suppose $P(AB,CD)$ is the cross ratio of a pencil of lines having vertex at P and $Q(EF,GH)$ is the cross ratio of a second pencil of lines having vertex at Q. If $P(AB,CD) = Q(EF,GH)$, is it true that $P(CA,BD) = Q(GE,FH)$? Explain.

11. Suppose the coordinates of a range of points A, B, C, D are 7, −5, 3, x, respectively. Use Definition 1 to determine the value of x such that $(AB,CD) = -3/8$.

CROSS RATIO AND HARMONIC SETS

12. Suppose the coordinates of a range of points A, B, C, D are -8, 9, -2, 4, respectively.
    (a) Use Definition 1 to determine the value of (AB,CD). (Recall that $m(P,Q) = q - p$, where p and q are the coordinates of P and Q, respectively.)
    (b) Use Theorem 7 to determine the value of each of (DA,CB), (BA,CD), (AC,DB), (CD,AB), and (BC,DA).

13. In **Figure 10.7**, A, B, C are any three points on line n. In addition, k is any second line through C, and E, F are points on k such that $\dfrac{m(C,E)}{m(C,F)} = \lambda$, for a given number λ. Line BF intersects line AE at point G. Line m contains G, is parallel to k, and intersects n at point D.
    Using the preceding information, develop a proof for this statement:
    For the points A, B, C, D, $(AB,CD) = \lambda$.

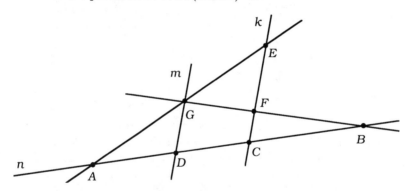

Figure 10.7

10.3 Harmonic Sets of Points and Lines

Harmonic sets are special sets of four points or four lines which have a particular relationship to each other based on relative positions. As in the preceding section, sets of points will be examined initially.

Definition 6: If A, B, C, D are four collinear points such that $(AB,CD) = -1$, then the points are called **harmonic points**, a **harmonic set**

of points, or a **harmonic range of points**.

Definition 7: If A, B, C, D are four harmonic points whose cross ratio is (AB,CD), the dividing points C and D of the segment (A,B) are called **harmonic conjugates** of the points A and B. Also, the dividing points A and B of the segment (C,D) are called **harmonic conjugates** of the points C and D.

Notation Rule 3: The pairing of four harmonic points A, B, C, D into conjugate pairs A, B and C, D will sometimes be denoted by $h(AB,CD)$.

Language Rule 1: The term **harmonic range** is generally employed with the notation $h(AB,CD)$.

Theorem 12: If four collinear points A, B, C, D form a harmonic range $h(AB,CD)$, then each of (AB,DC), (BA,CD), (BA,DC), (CD,AB), (CD,BA), (DC,AB), and (DC,BA) is a harmonic range.

Theorem 13: For four points A, B, C, D on a line n and a point P not on n, points A, B, C, D form a harmonic range $h(AB,CD)$ if and only if $m(E,B) = m(B,F)$, where E and F are the points where a line k parallel to line AP and containing B intersects lines PC and PD, respectively.

Theorem 14: If four points A, B, C, D forming a harmonic range $h(AB,CD)$ are centrally projected onto a second line, then A', B', C', D', the images of A, B, C, D, respectively, form a harmonic range $h(A'B',C'D')$.

Definition 8: A **harmonic pencil** is a pencil of four lines whose cross ratio is negative one (–1).

Notation Rule 4: The cross ratio of a harmonic pencil of lines having a point P for its vertex will sometimes be denoted as $hP(AB,CD)$, where (AB,CD) is the cross ratio of the range of points of intersection of these lines with a transversal.

Theorem 15: If a transversal intersects the lines of a pencil in a harmonic range of points, then every transversal of the pencil will intersect its lines in a harmonic range of points.

Exercises 10.3

1. Develop a proof for Theorem 12.

2. Develop a proof for Theorem 13.

3. Develop a proof for Theorem 14.

4. Develop a proof for Theorem 15.

5. In **Figure 10.8**, A, B, C are any three points on line n, with C not being the midpoint of segment of (A,B). In addition, k and p are any two parallel lines containing A and B, respectively, and q is any transversal containing C and intersecting k and p in points E and F, respectively. Point G is located on p such that $(F,B) \cong (B,G)$. Finally, line EG intersects n at the point D.
   Using the preceding information, develop a proof for this statement:
   The points A, B, C, D form a harmonic range $h(AB,CD)$.

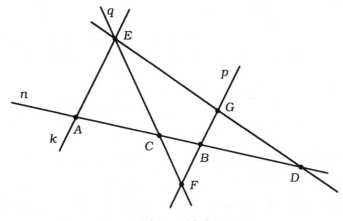

Figure 10.8

6. Suppose the coordinates of a range of points A, B, C, D are -12, 8, 12, x, respectively. Use Definition 6 to determine the value of x which will cause (AB,CD) to be a harmonic range.

10.4 Applications of Harmonic Sets

Harmonic sets occur in some unusual and unexpected places. The theorems in this section describe some of these occurrences.

Theorem 16: If the bisector of $\angle A$ of $\triangle ABC$ intersects side (B,C) in point D and the lines through B and C perpendicular to line AD intersect it at points E and F, respectively, then (AD,EF) is a harmonic range.

Definition 9: An **external bisector** of an angle is the bisector of an angle which forms a linear pair with the initial angle. (See **Figure 10.9**.)

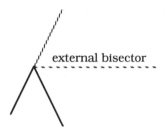

Figure 10.9

Language Rule 2: The bisector of an angle is sometimes called the **internal bisector** of the angle.

It should be observed that while an angle has only one internal bisector it has two external bisectors, which will, in fact, be collinear.

Theorem 17: If a point P is the vertex of an angle, then the sidelines of the angle and the lines containing the internal bisector and an external bisector of the angle form a harmonic pencil of lines having P for its vertex.

Proof: Suppose P is the vertex of an angle; further suppose that the sidelines of the angle and the lines containing the internal bisector and an external bisector of the angle are intersected by a transversal in the points A, B, C, D, respectively. (See **Figure 10.10**.)

By Appendices material, the point C divides side (A,B) of $\triangle ABP$ such that $\dfrac{m(A,C)}{m(C,B)} = \dfrac{m(P,A)}{m(P,B)}$. Also, by Appendices material, the point D divides side (A,B) of $\triangle ABP$ such that $\dfrac{m(A,D)}{m(B,D)} = \dfrac{m(P,A)}{m(P,B)}$.

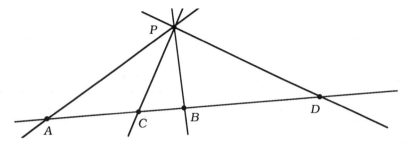

Figure 10.10

Thus, $\dfrac{m(A,C)}{m(C,B)} = \dfrac{m(A,D)}{m(B,D)}$; $\dfrac{m(A,C)}{m(C,B)} = -\dfrac{m(A,D)}{m(D,B)}$.

Hence, $\dfrac{m(A,C)/m(C,B)}{m(A,D)/m(D,B)} = -1$.

Therefore, $P(AB,CD) = -1$, and the four lines concurrent at P form a harmonic pencil. ∎

Theorem 18: If, for a circle, a chord AB is perpendicular to a diameter CD and P is a fifth point on the circle, then the lines PA, PB, PC, PD form a harmonic pencil.

Theorem 19: If square $ABCD$ is inscribed in a circle and P is a fifth point on the circle, then the lines PA, PB, PC, PD form a harmonic pencil.

Theorem 20: If, for a triangle $\triangle ABC$, the points D, E, F are the midpoints of sides (A,B), (B,C), (C,A), respectively, and lines DE and DC intersect line BF at points G, H, respectively, then (FG,HB) is a harmonic range.

Definition 10: For three positive numbers x, y, z, the number z is the **harmonic mean** of x and y if $\tfrac{1}{z}$ is the arithmetic mean of $\tfrac{1}{x}$ and $\tfrac{1}{y}$.

Theorem 21: Four collinear points A, B, C, D form a harmonic range $h(AB,CD)$ if and only if $m(A,B)$ is the harmonic mean of $m(A,C)$ and $m(A,D)$.

Proof: Suppose A, B, C, D are four collinear points.

Claim: If A, B, C, D form a harmonic range $h(AB,CD)$, then $m(A,B)$ is the harmonic mean of $m(A,C)$ and $m(A,D)$.

Proof: Suppose A, B, C, D form a harmonic range $h(AB,CD)$.
By Definition 6, $\dfrac{m(A,C)/m(C,B)}{m(A,D)/m(D,B)} = -1$. Thus, $\dfrac{m(A,C)}{m(C,B)} = -\dfrac{m(A,D)}{m(D,B)}$.

$$\dfrac{m(C,B)}{m(A,C)} = -\dfrac{m(D,B)}{m(A,D)}; \dfrac{m(C,B)}{m(A,C)} = \dfrac{m(B,D)}{m(A,D)}$$

$$\dfrac{1}{m(A,B)} \cdot \dfrac{m(C,B)}{m(A,C)} = \dfrac{1}{m(A,B)} \cdot \dfrac{m(B,D)}{m(A,D)}$$

$$\dfrac{m(C,B)}{m(A,B) \cdot m(A,C)} = \dfrac{m(B,D)}{m(A,B) \cdot m(A,D)}$$

$$\dfrac{m(A,B) - m(A,C)}{m(A,B) \cdot m(A,C)} = \dfrac{m(A,D) - m(A,B)}{m(A,B) \cdot m(A,D)},$$

since $m(C,B) = m(A,B) - m(A,C)$ and $m(B,D) = m(A,D) - m(A,B)$.

$$\dfrac{1}{m(A,C)} - \dfrac{1}{m(A,B)} = \dfrac{1}{m(A,B)} - \dfrac{1}{m(A,D)}$$

Therefore, $\dfrac{1}{m(A,C)} + \dfrac{1}{m(A,D)} = \dfrac{2}{m(A,B)}$.

Hence, $\dfrac{\dfrac{1}{m(A,C)} + \dfrac{1}{m(A,D)}}{2} = \dfrac{1}{m(A,B)}$, and $m(A,B)$ is the harmonic mean of $m(A,C)$ and $m(A,D)$. ◆

Claim: If $m(A,B)$ is the harmonic mean of $m(A,C)$ and $m(A,D)$, then A, B, C, D form a harmonic range $h(AB,CD)$.

Proof: Left as an exercise. ◆

Thus, A, B, C, D form a harmonic range $h(AB,CD)$ if and only if $m(A,B)$ is the harmonic mean of $m(A,C)$ and $m(A,D)$. ∎

It should be observed that the preceding theorem would be true if $m(A,C)$, $m(A,D)$ were replaced with $m(C,B)$, $m(D,B)$, respectively. Thus, $m(A,B)$ is the harmonic mean of $m(A,C)$, $m(A,D)$ and of $m(C,B)$, $m(D,B)$.

Definition 11: For three positive numbers x, y, z, the number z is the **mean proportional** of x and y if $x/z = z/y$ or, equivalently, $z^2 = xy$.

Theorem 22: Four collinear points A, B, C, D form a harmonic range $h(AB,CD)$ if and only if $m(K,B)$ is the mean proportional between $m(K,C)$ and $m(K,D)$, where K is the midpoint of segment (A,B).

Proof: Suppose A, B, C, D are four collinear points and K is the midpoint of (A,B).

Claim: If A, B, C, D form a harmonic range $h(AB,CD)$, then $m(K,B)$ is the mean proportional between $m(K,C)$ and $m(K,D)$.

Proof: Suppose A, B, C, D form a harmonic range $h(AB,CD)$.

By Definition 6, $\dfrac{m(A,C)/m(C,B)}{m(A,D)/m(D,B)} = -1$.

Thus, $\dfrac{m(A,C)}{m(C,B)} = -\dfrac{m(A,D)}{m(D,B)}$; $\dfrac{m(A,C)}{m(C,B)} = \dfrac{m(A,D)}{m(B,D)}$

$$\dfrac{m(A,K)+m(K,C)}{m(K,B)-m(K,C)} = \dfrac{m(A,K)+m(K,D)}{m(K,D)-m(K,B)},$$

since $m(A,C) = m(A,K)+m(K,C)$, $m(C,B) = m(K,B)-m(K,C)$, $m(A,D) = m(A,K)+m(K,D)$, and $m(B,D) = m(K,D)-m(K,B)$.

$\dfrac{m(K,B)+m(K,C)}{m(K,B)-m(K,C)} = \dfrac{m(K,B)+m(K,D)}{m(K,D)-m(K,B)}$, since $m(A,K) = m(K,B)$.

$$\left(m(K,B)+m(K,C)\right) \cdot \left(m(K,D)-m(K,B)\right)$$
$$= \left(m(K,B)+m(K,D)\right) \cdot \left(m(K,B)-m(K,C)\right);$$

$m(K,B) \cdot m(K,D) - (m(K,B))^2 + m(K,C) \cdot m(K,D) - m(K,C) \cdot m(K,B)$
$= (m(K,B))^2 - m(K,B) \cdot m(K,C) + m(K,D) \cdot m(K,B) - m(K,D) \cdot m(K,C);$

$$2m(K,C) \cdot m(K,D) = 2(m(K,B))^2.$$

Thus, $(m(K,B))^2 = m(K,C) \cdot m(K,D)$, and $m(K,B)$ is the mean proportional between $m(K,C)$ and $m(K,D)$ ♦

Claim: If $m(K,B)$ is the mean proportional between $m(K,C)$ and

$m(K,D)$, then A, B, C, D form a harmonic range $h(AB,CD)$.

Proof: Left as an exercise. ◆

Thus, A, B, C, D form a harmonic range $h(AB,CD)$ if and only if $m(K,B)$ is the mean proportional between $m(K,C)$ and $m(K,D)$, where K is the midpoint of (A,B). ■

It should be observed that the preceding theorem would be true if $m(K,B)$ were replaced with $m(A,K)$. Thus, each of $m(A,K)$ and $m(K,B)$ is the mean proportional between $m(K,C)$ and $m(K,D)$.

Definition 12: Two intersecting circles are called **orthogonal** if their tangent lines at a point of intersection are perpendicular.

Theorem 23: Two intersecting circles are orthogonal if and only if the line containing their centers intersects the circles in a harmonic range of points.

Proof: Suppose two circles having centers O and O', respectively, intersect at points P and Q; further suppose that line OO' intersects the circles in points A, B and C, D, respectively. (See **Figure 10.11**.)

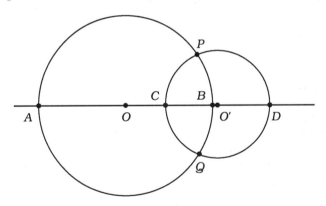

Figure 10.11

Claim: If the two circles are orthogonal, then A, B, C, D form a harmonic range of points.

Proof: Suppose the two circles are orthogonal.

Since the radius (O,P) of the circle centered at O is perpendicular to the line which is tangent to this circle at P, then line OP is tangent to the circle centered at O'. (See **Figure 10.12**.)

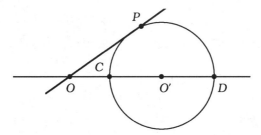

Figure 10.12

Thus, for the circle centered at O', OP is a tangent line and OD is a secant line. By Appendices material, $(m(O,P))^2 = m(O,C) \cdot m(O,D)$. Since each of (O,B) and (O,P) is a radius of the circle centered at O, $m(O,B) = m(O,P)$.

Hence, $(m(O,B))^2 = m(O,C) \cdot m(O,D)$.

Therefore, $m(O,B)$ is the mean proportional between $m(O,C)$ and $m(O,D)$, where O is the midpoint of segment (A,B).

By Theorem 22, (AB,CD) is a harmonic range. ◆

Claim: If (AB,CD) is a harmonic range, then the two circles are orthogonal.

Proof: Left as an exercise. ◆

Thus, two intersecting circles are orthogonal if and only if the line containing their centers intersects the circles in a harmonic range of points. ∎

The remaining theorems involving applications of harmonic sets are related to topics from Chapter 9. The proof of each can be developed using one or more of the theorems of Ceva, Gergonne, and Menelaus.

Theorem 24: If, for a triangle $\triangle ABC$, concurrent lines through vertices A, B, C intersect the opposite sides in points D, E, F, respectively, and the line FD intersects sideline AC at point G, then (CA,GE) is a harmonic range.

Proof: Suppose for $\triangle ABC$ concurrent lines through vertices A, B, C intersect the opposite sides in points D, E, F, respectively; further suppose line FD intersects sideline AC at point G. (See **Figure 10.13**.)

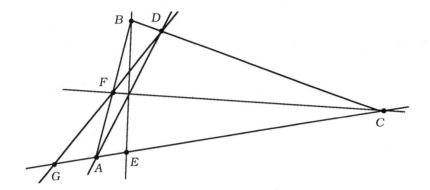

Figure 10.13

Since D, F, G are collinear points lying on the sidelines of $\triangle ABC$, then, by Menelaus' Theorem, $\dfrac{m(A,F)}{m(F,B)} \cdot \dfrac{m(B,D)}{m(D,C)} \cdot \dfrac{m(C,G)}{m(G,A)} = -1.$

Since concurrent lines AD, BE, CF are determined by the vertices and opposite side points of $\triangle ABC$, then, by Ceva's Theorem,

$$\frac{m(A,F)}{m(F,B)} \cdot \frac{m(B,D)}{m(D,C)} \cdot \frac{m(C,E)}{m(E,A)} = 1, \text{ or, equivalently}$$

$$\frac{m(F,B)}{m(A,F)} \cdot \frac{m(D,C)}{m(B,D)} \cdot \frac{m(E,A)}{m(C,E)} = 1.$$

Thus, $\left(\dfrac{m(A,F)}{m(F,B)} \cdot \dfrac{m(B,D)}{m(D,C)} \cdot \dfrac{m(C,G)}{m(G,A)}\right)\left(\dfrac{m(F,B)}{m(A,F)} \cdot \dfrac{m(D,C)}{m(B,D)} \cdot \dfrac{m(E,A)}{m(C,E)}\right) = -1.$

Hence, $\dfrac{m(C,G)}{m(G,A)} \cdot \dfrac{m(E,A)}{m(C,E)} = -1.$

Therefore, $\dfrac{m(C,G)/m(G,A)}{m(C,E)/m(E,A)} = -1$, and (CA,GE) is a harmonic range. ∎

Theorem 25: If, for a triangle $\triangle ABC$, its incircle is tangent to sides (A,B), (B,C), (C,A) at points E, F, G, respectively, and line EF intersects sideline AC at point D, then (AC,GD) is a harmonic range.

Theorem 26: If, for a triangle $\triangle ABC$, a line containing the midpoint D of side (B,C) intersects side (A,B) at E, sideline AC at F, and the line through A parallel to line BC at G, then (DG, EF) is a harmonic range.

Theorem 27: If, for a triangle $\triangle ABC$, points D, E, F are on sides (A,B), (B,C), (C,A), respectively, lines AE, BF, CD are concurrent, point G is the harmonic conjugate of D with respect to A and B, point H is the harmonic conjugate of E with respect to B and C, and point J is the harmonic conjugate of F with respect to A and C, then G, H, J are collinear points. (See **Figure 10.14**.)

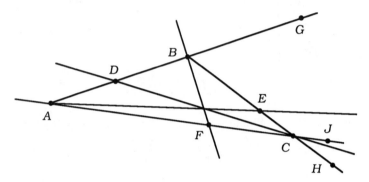

Figure 10.14

Exercises 10.4

1. Develop a proof for Theorem 16.

2. Develop a proof for Theorem 18.

3. Develop a proof for Theorem 19.

4. Develop a proof for Theorem 20.

5. Complete the proof for Theorem 21.

6. Complete the proof for Theorem 22.

7. Complete the proof for Theorem 23.

8. Develop a proof for Theorem 25.

9. Develop a proof for Theorem 26.

10. Develop a proof for Theorem 27.

****

SUPPLEMENTARY TOPICS

CONSTRUCTIONS

43. Construct a diagram similar to the one given below, and use constructions to determine the location of point D such that $(AB,CD) = 0.25$.

A B C

44. Construct a diagram similar to the one given above, and use constructions to determine the location of point D such that $(AB,CD) = -3$.

45. Construct a diagram similar to the one given below, and use constructions to determine a fourth line d containing point P such that the resulting pencil of lines will have a cross ratio of 2.

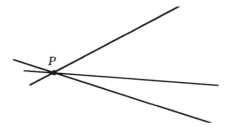

46. Construct a diagram similar to the one given below, and use constructions to determine two additional lines containing point P such that the resulting pencil of lines will be harmonic.

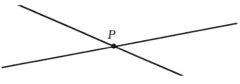

47. In the following diagram, points A and B are the endpoints of a diameter of the circle. Construct a diagram similar to this one, and use constructions to determine a second circle which contains point C and is orthogonal to the given circle.

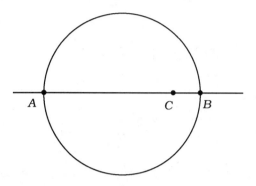

48. Construct a diagram similar to the one given below, and use constructions to determine the location of point D such that $m(A,B)$ is the harmonic mean of $m(C,B)$ and $m(D,B)$.

Chapter 11: Some Topics from Hyperbolic Geometry

11.1 Introduction

The most widely studied type of non-Euclidean geometry is hyperbolic geometry. It is generally believed that the German Carl Friedrich Gauss (1777-1855) initiated the idea of an axiom which allowed more than one parallel for a line from an external point. Unfortunately, he did not publish his work for fear of harming his reputation, for he was considered the outstanding mathematician of the early 1800's.

Working independently of each other and with no knowledge of Gauss' studies, the Russian Nicholas Lobachevsky (1793-1856) and the Hungarian János Bolyai (1802-1860) published results of their studies in 1829 and 1831, respectively. Each of these mathematicians based his work on a replacement for Euclid's parallel axiom which resulted in there existing two parallels for a line from an external point. This particular type of geometry was ultimately given the name hyperbolic geometry by the German mathematician Felix Klein (1849-1925).

If the contents of this chapter were to be a somewhat exhaustive study of hyperbolic geometry, it would begin with a negation of the parallel line existence axiom (Axiom 18 of Chapter 7) in which "exactly one" is replaced by "at least two." However, since the chapter will contain only an introduction to some topics from hyperbolic geometry, it is sufficient to begin with a stronger axiom, which is a modification of one first suggested by Hilbert.

Axiom 1: (Hyperbolic Parallel Axiom) If m is a line and A is a point not on m, then there exist exactly two noncollinear halflines $\overrightarrow{AB}$ and $\overrightarrow{AC}$ which do not intersect m and such that a third halfline $\overrightarrow{AD}$ intersects m if and only if $\overrightarrow{AD}$ is between $\overrightarrow{AB}$ and $\overrightarrow{AC}$.

Language Rule 1: In the preceding axiom, $\overrightarrow{AB}$ and $\overrightarrow{AC}$ can be called the **parallel halflines for m from A**. Also, each of lines AB and AC can be called a **parallel line for m through A**.

Definition 1: If $\overrightarrow{AB}$ and $\overrightarrow{AC}$ are the parallel halflines for a line m from a point A and D, E, F are points on m such that $D \bullet E \bullet F$, then each of $\overrightarrow{AB}$

and $\overrightarrow{AC}$ is **parallel** to the one of $\overrightarrow{ED}$ and $\overrightarrow{EF}$ lying on the same side of line AE. (See **Figure 11.1**.)

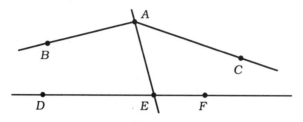

Figure 11.1

Definition 2: Two halflines which lie on the same side of the line containing their endpoints are said to have the **same direction**.

Since hyperbolic geometry differs from Euclidean and/or Hilbert geometry only in terms of the particular parallel axiom which is used, all the materials in Chapters 1 through 7 of Part II except those based on the parallel axiom can be employed in developing proofs for theorems in this chapter. In particular, all those materials except Axiom 18, Theorem 79, Theorem 80, Theorem 81, Theorem 82, Theorem 83, Theorem 93, and Theorem 94 can be used in this chapter.

11.2 Parallels

This section contains theorems which describe various characteristics of parallels in hyperbolic geometry. Some of these characteristics are analogous to those of parallels in Euclidean/Hilbert geometry, while others of them are quite different.

Theorem 1: If $\overrightarrow{AB}$ and $\overrightarrow{AC}$ are the parallel halflines for a line m from a point A and D is a point on m, then any halfline $\overrightarrow{AE}$ is between $\overrightarrow{AB}$ and $\overrightarrow{AD}$ or between $\overrightarrow{AD}$ and $\overrightarrow{AC}$ if and only if $\overrightarrow{AE}$ intersects m.

Proof: Suppose $\overrightarrow{AB}$ and $\overrightarrow{AC}$ are the parallel halflines for a line m from a point A, D is a point on m, and $\overrightarrow{AE}$ is a third halfline which does not contain D.

Since A is not on m, A and D are distinct and determine a line AD, one of whose halflines is $\overrightarrow{AD}$. Since $\overrightarrow{AD}$ intersects m, $\overrightarrow{AD}$ is between $\overrightarrow{AB}$ and $\overrightarrow{AC}$, by Axiom 1.

Claim: If $\overrightarrow{AE}$ is between $\overrightarrow{AB}$ and $\overrightarrow{AD}$, then $\overrightarrow{AE}$ intersects m.

Proof: Suppose $\overrightarrow{AE}$ is between $\overrightarrow{AB}$ and $\overrightarrow{AD}$. (See **Figure 11.2**.)

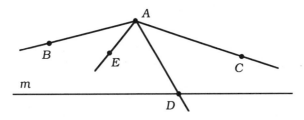

Figure 11.2

By Part II material (Theorem 45), $\overrightarrow{AE}$ is between $\overrightarrow{AB}$ and $\overrightarrow{AC}$.

Thus, by Axiom 1, $\overrightarrow{AE}$ intersects m. ♦

Claim: If $\overrightarrow{AE}$ is between $\overrightarrow{AD}$ and $\overrightarrow{AC}$, then $\overrightarrow{AE}$ intersects m.

Proof: Analogous to the preceding **Claim**. ♦

Claim: If $\overrightarrow{AE}$ intersects m, then $\overrightarrow{AE}$ is between $\overrightarrow{AB}$ and $\overrightarrow{AD}$ or between $\overrightarrow{AD}$ and $\overrightarrow{AC}$.

Proof: Suppose $\overrightarrow{AE}$ intersects m.

By Axiom 1, $\overrightarrow{AE}$ is between $\overrightarrow{AB}$ and $\overrightarrow{AC}$.

Since $\overrightarrow{AE}$ intersects m and does not contain D, then, by Part II material

(Theorem 2), $\vec{AE}$ and $\vec{AD}$ are disjoint.

By Part II material (Theorem 43), there exist points B^*, D^*, C^* on $\vec{AB}$, $\vec{AD}$, $\vec{AC}$, respectively, such that $B^{**}D^{**}C^*$.

Since $\vec{AE}$ is between $\vec{AB}$ and $\vec{AC}$ and, thus, lies in $int(\angle BAC)$, then $E \in int(\angle BAC)$. By Part II material (Theorem 41), $\vec{AE}$ intersects segment (B^*,C^*) at a point F.

Thus, F is a fourth point on line B^*C^* and $B^{**}D^{**}C^*$. By Part II material (Axiom 10), exactly one of $F \bullet B^{**}D^{**}C^*$, $B^{**}F \bullet D^{**}C^*$, $B^{**}D^{**}F \bullet C^*$, $B^{**}D^{**}C^{*} \bullet F$ is true.

Point F would not be in $int(\angle BAC)$ if $F \bullet B^{**}D^{**}C^*$ or $B^{**}D^{**}C^* \bullet F$ occurs. Thus, exactly one of $B^{**}F \bullet D^{**}C^*$ and $B^{**}D^{**}F \bullet C^*$ must occur.

If $B^{**}F \bullet D^{**}C^*$ or $B^{**}D^{**}F \bullet C^*$ occurs, then, by Part II material (Axiom 9), $B^{**}F \bullet D^*$ or $D^{**}F \bullet C^*$.

By Part II material (Theorem 38), $F \in int(\angle B^*AD^*)$ if $B^{**}F \bullet D^*$ and $F \in int(\angle D^*AC^*)$ if $D^{**}F \bullet C^*$.

Thus, by Part II material (Theorem 40), $\vec{AF} \subseteq int(\angle B^*AD^*)$ or $\vec{AF} \subseteq int(\angle D^*AC^*)$. Since $\angle BAD = \angle B^*AD^*$, $\angle DAC = \angle D^*AC^*$, and $\vec{AE} = \vec{AF}$, then $\vec{AE} \subseteq int(\angle BAD)$ or $\vec{AE} \subseteq int(\angle DAC)$.

Hence, $\vec{AE}$ is between $\vec{AB}$ and $\vec{AD}$ or $\vec{AE}$ is between $\vec{AD}$ and $\vec{AC}$. ♦

Therefore, if $\vec{AB}$ and $\vec{AC}$ are the parallel halflines for a line m from a point A and D is a point of m, then any halfline $\vec{AE}$ is between $\vec{AB}$ and $\vec{AD}$ or between $\vec{AD}$ and $\vec{AC}$ if and only if $\vec{AE}$ intersects m. ∎

Theorem 2: If $\vec{AB}$ and $\vec{AC}$ are the parallel halflines for a line m from a point A and D is a point on m, then $\vec{AD}$ is perpendicular to m if and only if $\angle BAD \cong \angle CAD$.

Proof: Suppose $\vec{AB}$ and $\vec{AC}$ are the parallel halflines for a line m from a point A and D is a point on m.

Since A is not on m, A and D are distinct and determine a line AD, one of whose halflines is $\vec{AD}$. Since $\vec{AD}$ intersects m, $\vec{AD}$ is between $\vec{AB}$ and $\vec{AC}$, by Axiom 1.

Claim: If $\vec{AD}$ is perpendicular to m, then $\angle BAD \cong \angle CAD$.

Proof: Suppose $\vec{AD}$ is perpendicular to m.

By Part II material (Theorem 64), exactly one of $\angle BAD < \angle CAD$, $\angle CAD < \angle BAD$ and $\angle BAD \cong \angle CAD$ can occur.

Subclaim: $\angle BAD < \angle CAD$ cannot occur.

Proof: Suppose $\angle BAD < \angle CAD$.

By Part II material, there exists a halfline $\vec{AE}$ such that $\vec{AE}$ is between $\vec{AC}$ and $\vec{AD}$ and $\angle BAD \cong \angle EAD$. (See **Figure 11.3**.)

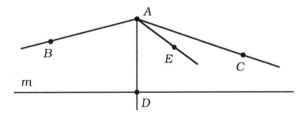

Figure 11.3

By Theorem 1, $\vec{AE}$ intersects m at a point F. By Part II material (Axiom 12), there exists a point G on m such that $G \bullet D \bullet F$ and $(G,D) \cong (D,F)$. (See **Figure 11.4**.)

From Part II material (Theorem 74, Theorem 70, and Axiom 13), $\angle ADG \cong \angle ADF$ and $(A,D) \cong (A,D)$. Hence, from Part II material (Theorem 50), $\triangle ADG \cong \triangle ADF$. Thus, $\angle GAD \cong \angle FAD$; since $\vec{AE} = \vec{AF}$, then $\angle GAD \cong \angle EAD$.

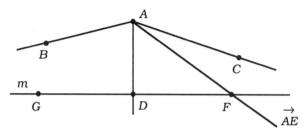

Figure 11.4

Since ∠BAD ≅ ∠EAD and ∠GAD ≅ ∠EAD, then, by Part II material (Axiom 16 and Axiom 15), ∠BAD ≅ ∠GAD and $\vec{AB} = \vec{AG}$.

Thus, $\vec{AB}$ intersects m at G. This contradicts the initial supposition that $\vec{AB}$ is parallel to m.

Hence, ∠BAD < ∠CAD cannot occur. ♦♦

Subclaim: ∠CAD < ∠BAD cannot occur.

Proof: Analogous to the preceding **Subclaim**. ♦♦

Therefore, if $\vec{AD}$ is perpendicular to m, then ∠BAD ≅ ∠CAD. ♦

Claim: If ∠BAD ≅ ∠CAD, then $\vec{AD}$ is perpendicular to m.

Proof: Suppose ∠BAD ≅ ∠CAD.

By Part II material (Theorem 90), there is a line which contains A, is perpendicular to m at a point E, and has $\vec{AE}$ as one of its halflines. By Axiom 1, $\vec{AE}$ is between $\vec{AB}$ and $\vec{AC}$, since it intersects m at E. (See **Figure 11.5**.)

From the first **Claim** in this proof, ∠BAE ≅ ∠CAE.

Since ∠BAD ≅ ∠CAD and ∠BAE ≅ ∠CAE, then each of $\vec{AD}$ and $\vec{AE}$ is a bisector of ∠BAC. By Part II material (Theorem 89), $\vec{AD} = \vec{AE}$.

Thus, $\vec{AD}$ is perpendicular to m. ♦

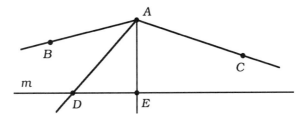

Figure 11.5

Therefore, if A is a point not on a line m, halflines $\overrightarrow{AB}$ and $\overrightarrow{AC}$ are parallels for m from A, and halfline $\overrightarrow{AD}$ intersects m at D, then $\overrightarrow{AD}$ is perpendicular to m if and only if $\angle BAD \cong \angle CAD$. ∎

Theorem 3: If $\overrightarrow{AB}$ and $\overrightarrow{AC}$ are the parallel halflines for a line m from point A and halfline $\overrightarrow{AD}$ is perpendicular to m at point D, then each of $\angle BAD$ and $\angle CAD$ is an acute angle.

Theorem 4: If A is a point not on a line m, halflines $\overrightarrow{AD}$ and $\overrightarrow{AE}$ are noncollinear with $\overrightarrow{AD}$ intersecting m at D and $\overrightarrow{AE}$ not intersecting m, and every halfline between $\overrightarrow{AD}$ and $\overrightarrow{AE}$ intersects m, then $\overrightarrow{AE}$ is parallel to m.

Proof: Suppose m is a line, A is a point not on m, halflines $\overrightarrow{AD}$ and $\overrightarrow{AE}$ are noncollinear with $\overrightarrow{AD}$ intersecting m at D and $\overrightarrow{AE}$ not intersecting m; suppose further that any halfline between $\overrightarrow{AD}$ and $\overrightarrow{AE}$ intersects m.

By Axiom 1, there exist two noncollinear halflines $\overrightarrow{AB}$ and $\overrightarrow{AC}$ which are parallel to m and such that $\overrightarrow{AD}$ is between $\overrightarrow{AB}$ and $\overrightarrow{AC}$. (See **Figure 11.6**.)

From Part II material (Theorem 41), $\overrightarrow{AD}$ intersects segment (B,C). Hence, B and C are on opposite sides of line AD.

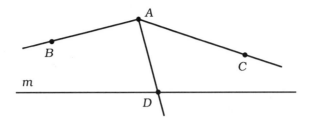

Figure 11.6

Since $\overrightarrow{AD}$ and $\overrightarrow{AE}$ are noncollinear, then E is not on line AD. Thus, E is on the B-side of AD or E is on the C-side of AD.

Claim: If E is on the B-side of AD, then $\overrightarrow{AE} = \overrightarrow{AB}$.

Proof: Suppose E is on the B-side of AD.

By Part II material (Theorem 21), all points of $\overrightarrow{AB}$ lie on the B-side of AD.

Since E is on the B-side of AD, then E is on $\overrightarrow{AB}$ or E, D are on the same side of $\overrightarrow{AB}$, or E, D are on opposite sides of $\overrightarrow{AB}$.

If E and D are on the same side of $\overrightarrow{AB}$, then E would be in $int(\angle BAD)$, causing $\overrightarrow{AE}$ to be between $\overrightarrow{AB}$ and $\overrightarrow{AD}$. By Theorem 1, $\overrightarrow{AE}$ would intersect m, which would contradict the initial supposition. Thus, E and D cannot be on the same side of $\overrightarrow{AB}$.

If E and D are on opposite sides of $\overrightarrow{AB}$, then $\overrightarrow{AB}$ would intersect segment (E,D). By Part II material (Theorem 38 and Theorem 40), $\overrightarrow{AB}$ would be between $\overrightarrow{AE}$ and $\overrightarrow{AD}$. By the initial supposition, $\overrightarrow{AB}$ would intersect m, which cannot occur because $\overrightarrow{AB}$ is parallel to m. Thus, E and D cannot be on opposite sides of $\overrightarrow{AB}$.

Thus, E is on $\overrightarrow{AB}$. Hence, $\overrightarrow{AE} = \overrightarrow{AB}$. ◆

Claim: If E is on the C-side of AD, then $\overrightarrow{AE} = \overrightarrow{AC}$.

Proof: Analogous to the preceding **Claim**. ♦

Thus, if A is a point not on a line m, halflines $\overrightarrow{AD}$ and $\overrightarrow{AE}$ are noncollinear with $\overrightarrow{AD}$ intersecting m at D and $\overrightarrow{AE}$ not intersecting m, and every halfline between $\overrightarrow{AD}$ and $\overrightarrow{AE}$ intersects m, then $\overrightarrow{AE}$ is parallel to m. ∎

Theorem 5: If $\overrightarrow{AB}$ is a parallel halfline for a line m from point A and $\overrightarrow{AK}$ is the opposite halfline for $\overrightarrow{AB}$, then $\overrightarrow{AK}$ is not parallel to m.

Theorem 6: If $\overrightarrow{AB}$ is a parallel halfline for a line m from point A and $\overrightarrow{AK}$ is the opposite halfline for $\overrightarrow{AB}$, then $\overrightarrow{AK}$ does not intersect m.

The statements in the two preceding theorems illustrate a major distinction between the plane geometry based on Hilbert's axioms, including a parallel axiom, and plane hyperbolic geometry. In Hilbert's geometry, if two lines are parallel, then *any* halfline lying on one of the lines is parallel to the other line. As a result of Theorem 5, this obviously is not true in hyperbolic geometry. What is true in hyperbolic geometry, however, is that if a halfline is parallel to a line then the line containing the halfline is also parallel to the line. In addition, Theorem 5 and Theorem 6 together indicate that in hyperbolic geometry there can exist a halfline which is neither parallel to nor intersecting with a line. This, too, is a situation not present in Hilbert's geometry.

Theorem 7: If $\overrightarrow{AB}$ is a parallel halfline for a line m from point A and K is a point of line AB such that $A \bullet K \bullet B$ or $K \bullet A \bullet B$, then halfline $\overrightarrow{KB}$ is parallel to m.

Proof: Suppose $\overrightarrow{AB}$ is a parallel halfline for a line m from point A and K is a point of line AB such that $A \bullet K \bullet B$ or $K \bullet A \bullet B$.

Claim: If $A \cdot K \cdot B$, then $\overrightarrow{KB}$ is parallel to m.

Proof: Suppose $A \cdot K \cdot B$.

By Part II material, $K \in \overrightarrow{AB}$, $\overrightarrow{KB} \subseteq \overrightarrow{AB}$, and $\overrightarrow{KB} \subseteq AB$. Thus, K is not on m and $\overrightarrow{KB}$ does not intersect m, since $\overrightarrow{AB}$ is parallel to m.

There exists at least one point E on m, by Part II material. Since $\overrightarrow{AB}$ is parallel to m and $E \in m$, then $E \notin \overrightarrow{AB}$. Thus, $K \neq E$. Therefore, there exists a line KE having $\overrightarrow{KE}$ as one of its halflines. (See **Figure 11.7**.)

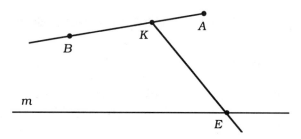

Figure 11.7

Since $E \in m$ and $\overrightarrow{AB}$ is parallel to m, then $E \notin \overrightarrow{AB}$. Thus, $E \notin \overrightarrow{KB}$, since $\overrightarrow{KB} \subseteq \overrightarrow{AB}$.

By Theorem 6, the opposite halfline for $\overrightarrow{AB}$ does not intersect m. Hence, E is not on the opposite halfline for $\overrightarrow{AB}$. Also, $E \neq A$, since A is not on m. Therefore, E is not on line AB. Thus, $\overrightarrow{KB}$ and $\overrightarrow{KE}$ are noncollinear.

Subclaim: Any halfline between $\overrightarrow{KB}$ and $\overrightarrow{KE}$ intersects m.

Proof: Left as an exercise. ◆◆

Thus, K is a point not on line m, $\overrightarrow{KB}$ and $\overrightarrow{KE}$ are noncollinear, $\overrightarrow{KE}$

intersects m at point E, $\overrightarrow{KB}$ does not intersect m, and any halfline between $\overrightarrow{KB}$ and $\overrightarrow{KE}$ intersects m. Hence, by Theorem 4, $\overrightarrow{KB}$ is parallel to m. ♦

Claim: If $K \bullet A \bullet B$, then $\overrightarrow{KB}$ is parallel to m.

Proof: Suppose $K \bullet A \bullet B$.
Since $K \bullet A \bullet B$, then, by Part II material (Axiom 6 and Axiom 1b), $K \in AB$, $AB = KB$, and $\overrightarrow{AK}$ is the opposite halfline for $\overrightarrow{AB}$. By Theorem 6, $\overrightarrow{AK}$ does not intersect m. Hence, K is not on m and no point of segment (A,K) lies on m. Thus, no point of $\overrightarrow{KB}$ lies on m, by Part II material, since $\overrightarrow{AB}$ is parallel to m and $A \notin m$.

There exists at least one point E on m, by Part II material. Since $E \in m$ and $K \notin m$, then $K \neq E$. Therefore, there exists a line KE having $\overrightarrow{KE}$ as one of its halflines. (See **Figure 11.8**.)

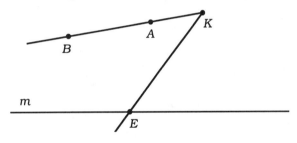

Figure 11.8

Since $A \notin m$ and no point of $\overrightarrow{AB}$ or $\overrightarrow{AK}$ is on m, then E is not on line AB. Thus, $\overrightarrow{KB}$ and $\overrightarrow{KE}$ are noncollinear.

Subclaim: Any halfline between $\overrightarrow{KB}$ and $\overrightarrow{KE}$ intersects m.

Proof: Suppose $\overrightarrow{KT}$ is a halfline between $\overrightarrow{KB}$ and $\overrightarrow{KE}$.
By Part II material, $\overrightarrow{KT} \subseteq int(\angle BKE)$, and, hence, $T \in int(\angle BKE)$. Thus,

by Part II material (Theorem 41), $\overrightarrow{KT}$ intersects segment (A,E) at a point Z. Since $Z \in \overrightarrow{KT}$, then $Z \in int(\angle BKE)$ and $\overrightarrow{KZ} = \overrightarrow{KT}$. By Part II material (Theorem 21), A and Z are on the same side of KE.

There exists a point F on line KZ such that $Z \bullet K \bullet F$, by Part II material (Axiom 5, Axiom 6, and Axiom 1b). Since K is on KE and KB, F and Z are on opposite sides of KE and on opposite sides of KB. (See **Figure 11.9**.) Thus, by Part II material (Theorem 22), A and F are on opposite sides of KE. Hence, by Part II material (Theorem 20), there exists a point X on KE such that $A \bullet X \bullet F$.

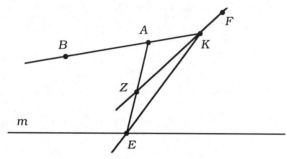

Figure 11.9

Since F is on one side of KB and $A \in KB$, then $F \neq A$. Thus, there exists a line AF, which, by Part II material (Axiom 5, Axiom 6, and Axiom 1b), contains a point G such that $G \bullet A \bullet F$. (See **Figure 11.10**.) By Part II material (Theorem 22), G and A are on the same side of KE.

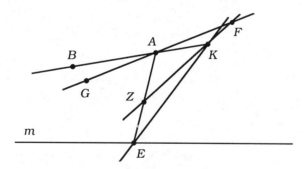

Figure 11.10

Since $G \bullet A \bullet F$ and $A \bullet X \bullet F$, where X is on KE, then, by Part II material (Theorem 6, Axiom 9, and Theorem 20), G and F are on opposite sides of KE.

Since A is on line KB and line AE, then G and F are on opposite sides of KB and on opposite sides of AE, by Part II material. Similarly, B and K are on opposite sides of AE. By Part II material (Theorem 21), F, K are on the same side of AE and E, Z are on the same side of KB.

By Part II material (Theorem 22), G and B are on the same side of AE. Thus, G is on the B-side of AE. Similarly, G and E are on the same side of KB. Since KB = AB, then G is on the E-side of AB. Hence, $G \in int(\angle BAE)$, and by Part II material (Theorem 40), $\overrightarrow{AG}$ is between $\overrightarrow{AB}$ and $\overrightarrow{AE}$.

Since $\overrightarrow{AB}$ is a parallel to m from A, $\overrightarrow{AE}$ intersects m at E, and $\overrightarrow{AG}$ is between $\overrightarrow{AB}$ and $\overrightarrow{AE}$, then, by Theorem 1, $\overrightarrow{AG}$ intersects m at a point H. Since $H \in \overrightarrow{AG}$, then $H \in AF$ and $AF = AH$. (See **Figure 11.11**.) By Part II material (Theorem 22), H and A are on the same side of KE. Similarly, H and Z are on the same side of KE.

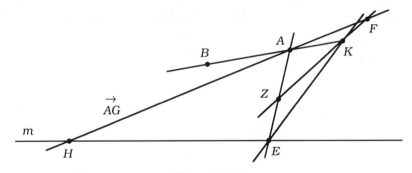

Figure 11.11

Line KZ cannot contain A, E, or H, by Part II material (Theorem 2). By Part II material (Axiom 11), KZ intersects (A,H) or (H,E). Since F is on AH and on KZ, KZ cannot intersect (A,H). Thus, KZ intersects (H,E) at a point J.

By Part II material (Theorem 21 and Theorem 22), J is on the H-side of KE, and, hence, on the Z-side of KE. Thus, $J \in \overrightarrow{KZ}$. Since $\overrightarrow{KZ} = \overrightarrow{KT}$, then $J \in \overrightarrow{KT}$.

Therefore, $\overrightarrow{KT}$ intersects m. ♦♦

Hence, K is a point not on line m, $\overrightarrow{KB}$ and $\overrightarrow{KE}$ are noncollinear, $\overrightarrow{KE}$ intersects m at point E, $\overrightarrow{KB}$ does not intersect m, and any halfline between $\overrightarrow{KB}$ and $\overrightarrow{KE}$ intersects m. Thus, by Theorem 4, $\overrightarrow{KB}$ is parallel to m. ◆

Therefore, if $\overrightarrow{AB}$ is a parallel halfline for a line m from point A and K is a point of line AB such that $A \bullet K \bullet B$ or $K \bullet A \bullet B$, then halfline $\overrightarrow{KB}$ is parallel to m. ∎

Theorem 8: If a pair of congruent alternate interior angles are formed when two halflines having the same direction are intersected by a transversal, then the halflines are not parallel.

Proof: Suppose two halflines $\overrightarrow{AB}$ and $\overrightarrow{CD}$ have the same direction and are intersected by transversal EF at points E and F, respectively. Suppose further that $\angle GEF \cong \angle CFE$, where $G \in \overrightarrow{AB}$ such that $A \bullet E \bullet G$. (See **Figure 11.12**.)

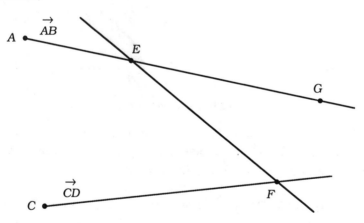

Figure 11.12

There are two possibilities:
(i) $\angle GEF$ is a right angle;
(ii) $\angle GEF$ is not a right angle.

Claim: If $\angle GEF$ is a right angle, then $\overrightarrow{AB}$ is not parallel to $\overrightarrow{CD}$.

Proof: Suppose $\angle GEF$ is a right angle.
By Part II material, $\angle GEF$ is not an acute angle.
Since $\angle GEF \cong \angle CFE$, then, by Part II material (Theorem 69), $\angle CFE$ is a right angle. Hence, $\overrightarrow{EF}$ is perpendicular to line CD at F and $\angle GEF$ is not an acute angle.

By Theorem 3, $\overrightarrow{EG}$ is not parallel to CD. By Theorem 7, $\overrightarrow{AE}$ is not parallel to CD. Since $\overrightarrow{AE} = \overrightarrow{AB}$, then $\overrightarrow{AB}$ is not parallel to CD.

Hence, by Definition 1, $\overrightarrow{AB}$ is not parallel to $\overrightarrow{CD}$. ♦

Claim: If $\angle GEF$ is not a right angle, then $\overrightarrow{AB}$ is not parallel to $\overrightarrow{CD}$.

Proof: Suppose $\angle GEF$ is not a right angle.
There are two possibilities:
(i) $\angle GEF$ is an acute angle;
(ii) $\angle GEF$ is an obtuse angle.

Subclaim: If $\angle GEF$ is an acute angle, then $\overrightarrow{AB}$ is not parallel to $\overrightarrow{CD}$.

Proof: Suppose $\angle GEF$ is an acute angle.
By Part II material (Theorem 88 and Theorem 21), segment (E,F) has a midpoint H which is not on AB or CD. By Part II material (Theorem 90), there exists a line which contains H and is perpendicular to AB at a point J.

Subsubclaim: $J \in \overrightarrow{EG}$.

Proof: Since $J \in AB$, and $A \bullet E \bullet G$, then, by Part II material (namely, Theorem 12), $J = E$, $J \in \overrightarrow{EA}$, or $J \in \overrightarrow{EG}$.
Since $\angle GEH$ is not a right angle, then $J \neq E$.

If $J \in \overrightarrow{EA}$, then, for $\triangle HEJ$, $\angle GEH$ is a remote exterior angle for the interior angle $\angle HJE$. Since $\angle HJE$ is a right angle and $\angle GEH$ is an acute angle, then $\angle HJE \not< \angle GEH$ and Theorem 84 of Part II would be contradicted. Thus, $J \notin \overrightarrow{EA}$.

Hence, $J \in \overrightarrow{EG}$. ♦♦♦

By Part II material (Axiom 12), there exists a point K on $\overrightarrow{FC}$ such that $(F,K) \cong (E,J)$. (See **Figure 11.13**.)

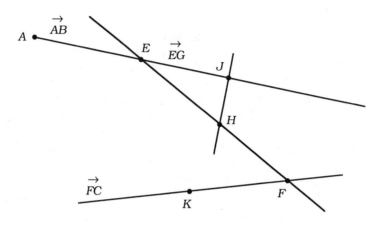

Figure 11.13

Since $H \in (E,F)$, $J \in \overrightarrow{EG}$, $K \in \overrightarrow{FC}$, then $\overrightarrow{EF} = \overrightarrow{EH}$, $\overrightarrow{FE} = \overrightarrow{FH}$, $\overrightarrow{EG} = \overrightarrow{EJ}$, and $\overrightarrow{FC} = \overrightarrow{FK}$. Thus, $\angle GEF = \angle JEH$ and $\angle CFE = \angle KFH$.

In $\triangle HEJ$ and $\triangle HFK$, $(H,E) \cong (H,F)$, $\angle JEH \cong \angle KFH$, and $(E,J) \cong (F,K)$. (See **Figure 11.14**.) Hence, by Part II material (Theorem 50), $\triangle HEJ \cong \triangle HFK$. Thus, $\angle EHJ \cong \angle FHK$ and $\angle HJE \cong \angle HKF$.

By Part II material (Axiom 5, Axiom 6, and Axiom 1b), $\overrightarrow{EJ}$ contains a point M such that $E \bullet J \bullet M$.

Since $\angle HJE \cong \angle HKF$ and $\angle HJE$ is a right angle, then, by Part II material (Theorem 69 and Theorem 74), each of $\angle HKF$ and $\angle MJH$ is a right angle.

Subsubclaim: $\overrightarrow{HJ}$ and $\overrightarrow{HK}$ are opposite halflines.

Proof: Angles $\angle EHJ$ and $\angle JHF$ form a linear pair of angles, since $\overrightarrow{HE}$

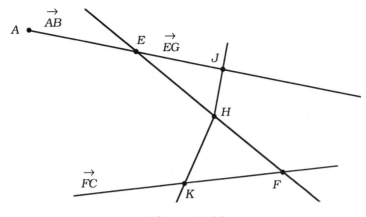

Figure 11.14

and $\overrightarrow{HF}$ are opposite halflines.

Since ∠EHJ ≅ ∠FHK and ∠JHF ≅ ∠JHF, then, by Part II material (Theorem 67), ∠FHK and ∠JHF are adjacent supplementary angles which form a linear pair. Thus, $\overrightarrow{HJ}$ and $\overrightarrow{HK}$ are opposite halflines. ♦♦♦

Since $K \in \overrightarrow{FC}$, then, by Part II material, exactly one of $K \bullet C \bullet F$, $K = C$, or $C \bullet K \bullet F$ is true.

Subsubclaim: If $K \bullet C \bullet F$, then $\overrightarrow{AB}$ is not parallel to $\overrightarrow{CD}$.

Proof: Suppose $K \bullet C \bullet F$.

By Part II material (Axiom 5), there exists a point P such that $P \bullet K \bullet C$. By Part II material (Theorem 6), $P \bullet K \bullet C \bullet F$. Thus by Part II material (Axiom 9, Axiom 6, and Axiom 1b), $K \in \overrightarrow{PF}$ and $\overrightarrow{CF} \subseteq \overrightarrow{PF}$. (See **Figure 11.15**.)

Since ∠HKF is a right angle, then, by Part II material (Theorem 74 and Theorem 69), ∠PKH ≅ ∠MJH. Thus, ∠PKJ ≅ ∠MJK, since $\overrightarrow{KH} = \overrightarrow{KJ}$ and $\overrightarrow{JH} = \overrightarrow{JK}$.

Thus, $\overrightarrow{AB}$ and $\overrightarrow{PF}$ have the same direction and are intersected by transversal JK at points J and K, respectively, ∠MJK ≅ ∠PKJ, and ∠MJK is a right angle.

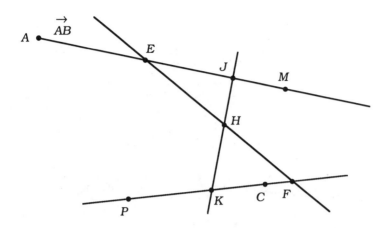

Figure 11.15

By the initial **Claim**, $\vec{AB}$ is not parallel to $\vec{PF}$.

By Theorem 7, $\vec{AB}$ is not parallel to $\vec{CF}$. Since $F \in \vec{CD}$, then $\vec{CD} = \vec{CF}$ and $\vec{AB}$ is not parallel to $\vec{CD}$. ♦♦♦

Subsubclaim: If $K = C$, then $\vec{AB}$ is not parallel to $\vec{CD}$.

Proof: Left as an exercise. ♦♦♦

Subsubclaim: If $C \bullet K \bullet F$, then $\vec{AB}$ is not parallel to $\vec{CD}$.

Proof: Left as an exercise. ♦♦♦

Thus, if $\angle GEF$ is an acute angle, then $\vec{AB}$ is not parallel to $\vec{CD}$. ♦♦

Subclaim: If $\angle GEF$ is an obtuse angle, then $\vec{AB}$ is not parallel to $\vec{CD}$.

Proof: Left as an exercise. ♦♦

Hence, if $\angle GEF$ is not a right angle, then $\vec{AB}$ is not parallel to $\vec{CD}$. ♦

Thus, if a pair of congruent alternate interior angles are formed when two halflines having the same direction are intersected by a transversal, then the halflines are not parallel. ∎

Theorem 9: If a pair of congruent corresponding angles are formed when two halflines having the same direction are intersected by a transversal, then the halflines are not parallel.

Theorem 10: If one halfline is parallel to a second halfline, then the second halfline is parallel to the first halfline.

Proof: Suppose halfline $\overrightarrow{AB}$ is parallel to halfline $\overrightarrow{EF}$.

By Definition 1, $\overrightarrow{AB}$ is a parallel halfline for a line EF from A and $\overrightarrow{EF}$ is the halfline of EF which lies on the same side of line AE as does $\overrightarrow{AB}$.

By Part II material (Theorem 90), there exists a line AG which is perpendicular to line EF at point G and has $\overrightarrow{AG}$ as one of its halflines. (See **Figure 11.16.**)

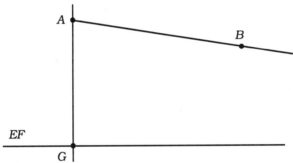

Figure 11.16

There are five possibilities for the location of G: $G \bullet E \bullet F$, $G = E$, $E \bullet G \bullet F$, $G = F$, $E \bullet F \bullet G$.

Claim: If $G \bullet E \bullet F$, then $\overrightarrow{EF}$ is parallel to $\overrightarrow{AB}$.

Proof: Suppose $G \bullet E \bullet F$. (See **Figure 11.17.**)

Since $\overrightarrow{AG}$ is perpendicular to line EF at G, then $\angle AGE$ is a right angle.

SOME TOPICS FROM HYPERBOLIC GEOMETRY 265

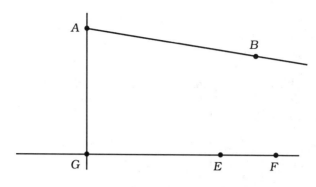

Figure 11.17

By Axiom 1, $\overrightarrow{AB}$ does not intersect EF and A is not on EF. By Theorem 6, the opposite halfline for $\overrightarrow{AB}$ on line AB does not intersect EF. Hence, $\overrightarrow{GE}$ does not intersect AB. Also, $\overrightarrow{GE}$ and $\overrightarrow{GA}$ are noncollinear, with $\overrightarrow{GA}$ intersecting AB at A.

Subclaim: Any halfline between $\overrightarrow{GE}$ and $\overrightarrow{GA}$ intersects AB.

Proof: Suppose $\overrightarrow{GH}$ is a halfline between $\overrightarrow{GE}$ and $\overrightarrow{GA}$. (See **Figure 11.18**.)

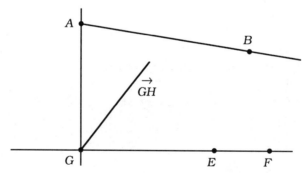

Figure 11.18

By Part II material, $\angle AGH < \angle AGE$.

By Part II material (Theorem 90), there exists line AJ which is perpendicular to GH at point J and has $\overrightarrow{AJ}$ as one of its halflines.

Subsubclaim: $J \in \overrightarrow{GH}$.

Proof: Left as an exercise. ♦♦♦

By Part II material (Theorem 64), exactly one of $\angle GAB < \angle GAJ$, $\angle GAB \cong \angle GAJ$, and $\angle GAJ < \angle GAB$ is true.

Subsubclaim: If $\angle GAB < \angle GAJ$, then $\overrightarrow{GH}$ intersects AB.

Proof: Left as an exercise. ♦♦♦

Subsubclaim: If $\angle GAB \cong \angle GAJ$, then $\overrightarrow{GH}$ intersects AB.

Proof: Left as an exercise. ♦♦♦

Subsubclaim: If $\angle GAJ < \angle GAB$, then $\overrightarrow{GH}$ intersects AB.

Proof: Suppose $\angle GAJ < \angle GAB$.

By Part II material, $\overrightarrow{AJ}$ is between $\overrightarrow{AG}$ and $\overrightarrow{AB}$. Thus, $\angle BAJ < \angle GAB$.

By Part II material, there exists a halfline $\overrightarrow{AM}$ such that $\overrightarrow{AM}$ is between $\overrightarrow{AG}$ and $\overrightarrow{AB}$ and $\angle MAG \cong \angle BAJ$. By Theorem 1, $\overrightarrow{AM}$ intersects EF at a point N. By Part II material (Theorem 21 and Theorem 22), $N \in \overrightarrow{GE}$. (See **Figure 11.19**.) Thus, $\overrightarrow{AM} = \overrightarrow{AN}$, $\angle MAG = \angle NAG$, and $\angle NAG \cong \angle BAJ$. Also, $\overrightarrow{GN} = \overrightarrow{GE}$, and $\angle AGN = \angle AGE$.

In $\triangle AGJ$, $\angle AGJ$ is an acute angle since $\overrightarrow{GJ}$ is between $\overrightarrow{GA}$ and $\overrightarrow{GN}$ and $\angle AGN$ is a right angle. Hence, by Part II material, $\angle AGJ < \angle AJG$, since $\angle AJG$ is a right angle.

Thus, by Part II material (Theorem 86), $(A,J) < (A,G)$. By Part II material, there exists a point P such that $A \bullet P \bullet G$ and $(A,P) \cong (A,J)$.

By Part II material (Theorem 92), there exists a line m which is perpendicular to $\overrightarrow{AG}$ at P. (See **Figure 11.20**.)

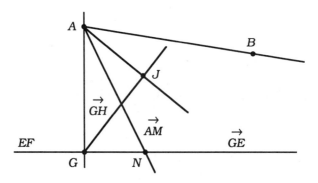

Figure 11.19

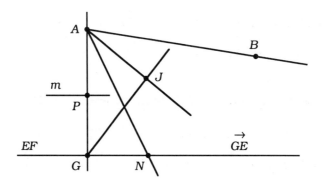

Figure 11.20

By Part II material (Theorem 78), m does not intersect EF. Thus, by Part II material (Axiom 11), m intersects side (A,N) of $\triangle AGN$ in a point Q. (See **Figure 11.21**.) By Part II material (Axiom 12), there exists a point R such that $G \bullet J \bullet R$ and $(J,R) \cong (P,Q)$. By Part II material (Theorem 74), $\angle AJR$ is a right angle. By Part II material (Theorem 70), $\angle APQ \cong \angle AJR$, since each is a right angle.

In $\triangle APQ$ and $\triangle AJR$, $(A,P) \cong (A,J)$, $\angle APQ \cong \angle AJR$, and $(P,Q) \cong (J,R)$. (See **Figure 11.22**.) Hence, by Part II material (Theorem 50), $\triangle QAP \cong \triangle RAJ$. Thus, $\angle QAP \cong \angle RAJ$.

Since $\overrightarrow{AQ} = \overrightarrow{AN}$, $\overrightarrow{AP} = \overrightarrow{AG}$, and $\angle QAP = \angle NAG$, then $\angle RAJ \cong \angle NAG$.

Thus, $\angle RAJ \cong \angle NAG$ and $\angle NAG \cong \angle BAJ$. By Part II material (Axiom 16), $\angle RAJ \cong \angle BAJ$.

By Part II material (Axiom 15), $\overrightarrow{AR} = \overrightarrow{AB}$. Hence, $\overrightarrow{GJ}$ intersects $\overrightarrow{AB}$ at

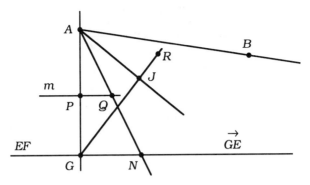

Figure 11.21

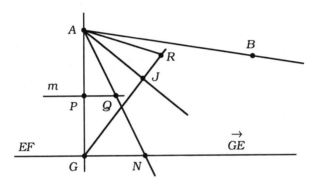

Figure 11.22

R. Since $\vec{GJ} = \vec{GH}$, then $\vec{GH}$ intersects AB at R. ♦♦♦

Therefore, if $G \bullet E \bullet F$, then any halfline between $\vec{GE}$ and $\vec{GA}$ intersects AB. ♦♦

By Theorem 4, $\vec{GE}$ is parallel to AB. Since $G \bullet E \bullet F$, then $\vec{EF}$ is parallel to AB, by Theorem 7. Thus, by Definition 1, $\vec{EF}$ is parallel to $\vec{AB}$. ♦

Claim: If $G = E$, then $\vec{EF}$ is parallel to $\vec{AB}$.

Proof: Left as an exercise. ♦

SOME TOPICS FROM HYPERBOLIC GEOMETRY 269

Claim: If $E \bullet G \bullet F$, then $\overrightarrow{EF}$ is parallel to $\overrightarrow{AB}$.

Proof: Left as an exercise. ◆

Claim: If $G = F$, then $\overrightarrow{EF}$ is parallel to $\overrightarrow{AB}$.

Proof: Left as an exercise. ◆

Claim: If $E \bullet F \bullet G$, then $\overrightarrow{EF}$ is parallel to $\overrightarrow{AB}$.

Proof: Left as an exercise. ◆

Therefore, halfline $\overrightarrow{EF}$ is parallel to halfline $\overrightarrow{AB}$. ∎

Theorem 11: If each of two halflines is parallel to a third halfline, then they are parallel to each other.

Proof: Suppose each of two halflines $\overrightarrow{AB}$ and $\overrightarrow{CD}$ is parallel to a third halfline $\overrightarrow{EF}$.

By Theorem 10, $\overrightarrow{EF}$ is parallel to each of $\overrightarrow{AB}$ and $\overrightarrow{CD}$.

By Definition 1, $\overrightarrow{AB}$ is a parallel halfline for line EF from A and $\overrightarrow{EF}$ is the halfline of EF which lies on the same side of line AE as does $\overrightarrow{AB}$, $\overrightarrow{EF}$ is a parallel halfline for line AB from E and $\overrightarrow{AB}$ is the halfline of AB which lies on the same side of AE as does $\overrightarrow{EF}$, $\overrightarrow{CD}$ is a parallel halfline for line EF from C and $\overrightarrow{EF}$ is the halfline of EF which lies on the same side of line CE as does $\overrightarrow{CD}$, and $\overrightarrow{EF}$ is a parallel halfline for line CD from E and $\overrightarrow{CD}$ is the halfline of CD which lies on the same side of CE as does $\overrightarrow{EF}$.

There are two possible locations for $\vec{EF}$:

(i) $\vec{AB}$ and $\vec{CD}$ are on opposite sides of $\vec{EF}$;

(ii) $\vec{AB}$ and $\vec{CD}$ are on the same side of $\vec{EF}$.

Claim: If $\vec{AB}$ and $\vec{CD}$ are on opposite sides of $\vec{EF}$, then $\vec{AB}$ and $\vec{CD}$ are parallel to each other.

Proof: Suppose $\vec{AB}$ and $\vec{CD}$ are on opposite sides of $\vec{EF}$.

By Theorem 6, the opposite halfline for $\vec{AB}$ does not intersect line EF and the opposite halfline for $\vec{CD}$ does not intersect EF. Hence, by Part II material (Theorem 20 and Theorem 22), lines AB and CD are on opposite sides of line EF. Thus, AB and CD are disjoint.

Since A and C are on opposite sides of EF, then, by Part II material (Theorem 20), segment (A,C) intersects EF at a point G. (See **Figure 11.23**.)

Also, $\vec{AB}$ and $\vec{AC}$ are noncollinear.

By Part II material (Theorem 20, Theorem 21, and Theorem 22), $\vec{AB}$ and $\vec{CD}$ are on the same side of AC.

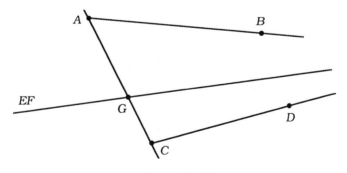

Figure 11.23

There are five possible locations for G: $G \bullet E \bullet F$, $G = E$, $E \bullet G \bullet F$, $G = F$, $E \bullet F \bullet G$.

Subclaim: If $G \bullet E \bullet F$, then any halfline between $\overrightarrow{AB}$ and $\overrightarrow{AC}$ intersects CD.

Proof: Left as an exercise. ♦♦

Subclaim: If $G = E$, then any halfline between $\overrightarrow{AB}$ and $\overrightarrow{AC}$ intersects CD.

Proof: Left as an exercise. ♦♦

Subclaim: If $E \bullet G \bullet F$, then any halfline between $\overrightarrow{AB}$ and $\overrightarrow{AC}$ intersects CD.

Proof: Suppose $E \bullet G \bullet F$ and $\overrightarrow{AH}$ is a halfline between $\overrightarrow{AB}$ and $\overrightarrow{AC}$.

By Theorem 7 and Theorem 10, $\overrightarrow{AB}$, $\overrightarrow{GF}$ are parallel to each other and $\overrightarrow{CD}$, $\overrightarrow{GF}$ are parallel to each other.

Since $\overrightarrow{AB}$ is parallel to EF from A and $\overrightarrow{AC}$ intersects EF at G, then, by Theorem 1, $\overrightarrow{AH}$ intersects EF at a point J. (See **Figure 11.24**.) Since $J \in \overrightarrow{AH}$, then $\overrightarrow{AJ} = \overrightarrow{AH}$.

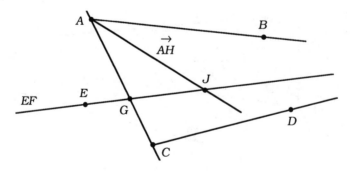

Figure 11.24

By Part II material (namely, Theorem 21), $J \in \overrightarrow{GF}$.

There are three possible locations for J: $G \bullet J \bullet F$, $J = F$, $G \bullet F \bullet J$.

Subsubclaim: If $G \bullet J \bullet F$, then $\overrightarrow{AH}$ intersects CD.

Proof: Suppose $G \bullet J \bullet F$.

By Theorem 7, $\overrightarrow{JF}$ is parallel to $\overrightarrow{CD}$. Hence, $J \neq C$, and there exists line CJ having $\overrightarrow{JC}$ as one of its halflines.

By Part II material (Axiom 5, Axiom 6, and Axiom 1b), there exists a point K on $\overrightarrow{AJ}$ such that $A \bullet J \bullet K$. (See **Figure 11.25**.)

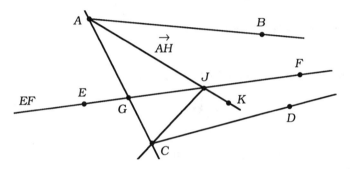

Figure 11.25

Since $J \in EF$ and $J \in CJ$, then A, K are on opposite sides of EF, A, K are on opposite sides of CJ, and G, F are on opposite sides of CJ. By Part II material (Theorem 21), A and G are on the same side of CJ. Hence, by Part II material (Theorem 22), K is on the F-side of CJ and K is on the C-side of EF. Since $JF = EF$, then K is on the C-side of JF. Thus, $K \in int(\angle CJF)$ and, by Part II material (Theorem 40), $\overrightarrow{JK}$ is between $\overrightarrow{JF}$ and $\overrightarrow{JC}$.

Since $\overrightarrow{JF}$ is parallel to $\overrightarrow{CD}$, then, by Definition 1, $\overrightarrow{JF}$ is parallel to CD from J. Since $\overrightarrow{JF}$ is parallel to CD from J, $\overrightarrow{JC}$ intersects CD at C, and $\overrightarrow{JK}$ is between $\overrightarrow{JF}$ and $\overrightarrow{JC}$, then $\overrightarrow{JK}$ intersects CD at a point M, by Theorem 1. (See **Figure 11.26**.)

Hence, $\overrightarrow{AH}$ intersects CD at M, since $\overrightarrow{JK} \subseteq \overrightarrow{AJ}$ and $\overrightarrow{AJ} = \overrightarrow{AH}$. ♦♦♦

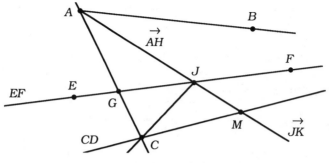

Figure 11.26

Subsubclaim: If $J = F$, then $\overrightarrow{AH}$ intersects CD.

Proof: Left as an exercise. ♦♦♦

Subsubclaim: If $G \bullet F \bullet J$, then $\overrightarrow{AH}$ intersects CD.

Proof: Left as an exercise. ♦♦♦

Therefore, if $E \bullet G \bullet F$, then any halfline between $\overrightarrow{AB}$ and $\overrightarrow{AC}$ intersects CD. ♦♦

Subclaim: If $G = F$, then any halfline between $\overrightarrow{AB}$ and $\overrightarrow{AC}$ intersects CD.

Proof: Left as an exercise. ♦♦

Subclaim: If $E \bullet F \bullet G$, then any halfline between $\overrightarrow{AB}$ and $\overrightarrow{AC}$ intersects CD.

Proof: Left as an exercise. ♦♦

Therefore, since A is not on CD, halflines $\overrightarrow{AB}$ and $\overrightarrow{AC}$ are noncollinear, $\overrightarrow{AC}$ intersects CD at C, $\overrightarrow{AB}$ does not intersect CD, and any halfline between

$\overrightarrow{AB}$ and $\overrightarrow{AC}$ intersects CD, then $\overrightarrow{AB}$ is parallel to CD.

By Definition 1, $\overrightarrow{AB}$ is parallel to $\overrightarrow{CD}$, and, by Theorem 10, $\overrightarrow{CD}$ is parallel to $\overrightarrow{AB}$.

Thus, if $\overrightarrow{AB}$ and $\overrightarrow{CD}$ are on opposite sides of $\overrightarrow{EF}$, then $\overrightarrow{AB}$ and $\overrightarrow{CD}$ are parallel to each other. ◆

Claim: If $\overrightarrow{AB}$ and $\overrightarrow{CD}$ are on the same side of $\overrightarrow{EF}$, then $\overrightarrow{AB}$ and $\overrightarrow{CD}$ are parallel to each other.

Proof: Suppose $\overrightarrow{AB}$ and $\overrightarrow{CD}$ are on the same side of $\overrightarrow{EF}$.

Subclaim: $\overrightarrow{AB}$ and $\overrightarrow{CD}$ do not intersect.

Proof: Left as an exercise. ◆◆

Since $\overrightarrow{AB}$ and $\overrightarrow{CD}$ do not intersect, then $\overrightarrow{AB}$, $\overrightarrow{CD}$, $\overrightarrow{EF}$ are three nonintersecting halflines, with $\overrightarrow{AB}$, $\overrightarrow{CD}$ lying on the same side of $\overrightarrow{EF}$. Hence, $\overrightarrow{AB}$, $\overrightarrow{EF}$ are on opposite sides of $\overrightarrow{CD}$, or $\overrightarrow{CD}$, $\overrightarrow{EF}$ are on opposite sides of $\overrightarrow{AB}$.

Subclaim: If $\overrightarrow{AB}$ and $\overrightarrow{EF}$ are on opposite sides of $\overrightarrow{CD}$, then $\overrightarrow{AB}$ and $\overrightarrow{CD}$ are parallel to each other.

Proof: Suppose $\overrightarrow{AB}$ and $\overrightarrow{EF}$ are on opposite sides of $\overrightarrow{CD}$.
By Part II material (Axiom 5), there exists a point G such that $A \bullet G \bullet B$.
By Part II material (Axiom 6 and Axiom 1b), $G \in \overrightarrow{AB}$.

By Axiom 1 and Definition 1, there exists exactly one halfline $\overrightarrow{GH}$ which is parallel to $\overrightarrow{CD}$ and lies on the same side of line GC as does $\overrightarrow{CD}$. (See

Figure 11.27.) By Theorem 10, $\vec{CD}$ is parallel to $\vec{GH}$.

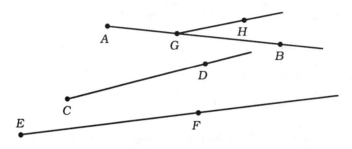

Figure 11.27

Since $G \in \vec{AB}$, $\vec{AB}$ and $\vec{EF}$ are on opposite sides of $\vec{CD}$, and $\vec{GH}$ is parallel to $\vec{CD}$, then, by the preceding **Subclaim** and Part II material (Theorem 20), $\vec{GH}$ and $\vec{EF}$ are on opposite sides of $\vec{CD}$. Since $\vec{GH}$ and $\vec{EF}$ are both parallel to $\vec{CD}$, then, by the initial **Claim**, $\vec{GH}$ and $\vec{EF}$ are parallel to each other.

By Theorem 7, $\vec{GB}$ is parallel to $\vec{EF}$, since $A \bullet G \bullet B$ and $\vec{AB}$ is parallel to $\vec{EF}$.

Thus, from point G there are halflines $\vec{GH}$ and $\vec{GB}$ which are parallel to $\vec{EF}$ and lying on the same side of line GE. By Axiom 1 and Definition 1, $\vec{GH} = \vec{GB}$.

Hence, $\vec{GB}$ is parallel to $\vec{CD}$. By Theorem 7, $\vec{AB}$ is parallel to $\vec{CD}$, and, by Theorem 10, $\vec{CD}$ is parallel to $\vec{AB}$.

Therefore, $\vec{AB}$ and $\vec{CD}$ are parallel to each other. ♦♦

Subclaim: If $\vec{CD}$ and $\vec{EF}$ are on opposite sides of $\vec{AB}$, then $\vec{AB}$ and $\vec{CD}$ are parallel to each other.

Proof: Analogous to the preceding **Subclaim**. ♦♦

Thus, if $\vec{AB}$ and $\vec{CD}$ are on the same side of $\vec{EF}$, then $\vec{AB}$ and $\vec{CD}$ are parallel to each other. ♦

Therefore, if each of two halflines $\vec{AB}$ and $\vec{CD}$ is parallel to a third halfline $\vec{EF}$, then $\vec{AB}$ and $\vec{CD}$ are parallel to each other. ∎

Exercises 11.2

1. Develop a proof for Theorem 3.

2. Develop a proof for Theorem 5.

3. Develop a proof for Theorem 6.

4. Complete the proof for Theorem 7.

5. Complete the proof for Theorem 8.

6. Develop a proof for Theorem 9.

7. Complete the proof for Theorem 10.

8. Complete the proof for Theorem 11.

11.3 Trilaterals

The concept to be examined in this section has no counterpart in Euclidean/Hilbert plane geometry. However, as the theorems will indicate, many of its characteristics are the same as those of the Euclidean/Hilbert triangle.

Definition 3: If $\vec{XA}$ and $\vec{YD}$ are two parallel halflines intersected by a transversal at points B and C, respectively, where $X \bullet B \bullet A$ and $Y \bullet C \bullet D$, then the set which is the union of $\{B,C\}$, segment (B,C), halfline $\vec{BA}$, and halfline $\overrightarrow{CD}$ is called a **trilateral**. (See **Figure 11.28**.) Points B and C are called the **vertices**, segment (B,C) is called the **inner side**, and halflines $\overrightarrow{BA}$ and $\overrightarrow{CD}$ are called the **outer sides**. Angles $\angle ABC$ and $\angle DCB$ are called the **angles**, and angles $\angle XBC$ and $\angle YCB$ and their respective vertical angles are called the **exterior angles**.

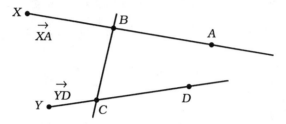

Figure 11.28

Notation Rule 1: If a trilateral is the union of $\{B,C\}$, segment (B,C), halfline $\overrightarrow{BA}$, and halfline $\overrightarrow{CD}$, then it may be denoted as **T**ABCD. In the symbol **T**ABCD, points B and C are the vertices of the trilateral.

Language Rule 2: The inner side and two outer sides of a trilateral are sometimes called the **sides** of the trilateral.

Theorem 12: The outer sides of a trilateral are parallel.

Definition 4: The **interior** of a trilateral **T**ABCD is the intersection of:
(a)   the side of line BC which contains A and D;
(b)   the side of line BA which contains C and D; and
(c)   the side of line CD which contains B and A.

Notation Rule 2: The interior of a trilateral **T**ABCD will be denoted as $int(\mathbf{T}ABCD)$.

Definition 5: The **exterior** of a trilateral **T**ABCD is the set of all points which are neither on **T**ABCD nor in its interior.

Theorem 13: The bisectors of the angles of a trilateral intersect in a point which is in the interior of the trilateral.

Theorem 14: The interior of a trilateral is nonempty.

Definition 6: In a trilateral, the vertex which is not the endpoint for a particular outer side is called the **opposite vertex** for that outer side, while the outer side is called the **opposite outer side** for that vertex.

Theorem 15: If a line passes through a vertex of a trilateral and contains a point in the trilateral's interior, then it intersects the outer side opposite that vertex.

Proof: Suppose for trilateral $TABCD$ point P is in the interior of $TABCD$ and line m contains P and a vertex of $TABCD$; without loss of generality, suppose m contains vertex B. (See **Figure 11.29**.)

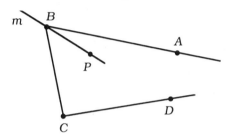

Figure 11.29

Since B and P are on m, then $\overrightarrow{BP}$ is a halfline of m. Since $P \in int(TABCD)$, then P lies on the side of AB which contains C and P lies on the side of BC which contains A. Hence, $P \in int(\angle ABC)$, and, by Part II material (Theorem 40), $\overrightarrow{BP} \subseteq int(\angle ABC)$. Thus, $\overrightarrow{BP}$ is between $\overrightarrow{BA}$ and $\overrightarrow{BC}$.

By Theorem 12, $\overrightarrow{BA}$ is parallel to $\overrightarrow{CD}$. By Definition 1, $\overrightarrow{BA}$ is parallel to CD from B.

Hence, $\overrightarrow{BA}$ is parallel to line CD from point B, $\overrightarrow{BC}$ intersects CD at C, and $\overrightarrow{BP}$ is between $\overrightarrow{BA}$ and $\overrightarrow{BC}$. Therefore, by Theorem 1, $\overrightarrow{BP}$ intersects CD.

By Part II material (Theorem 21 and Theorem 22), $\overrightarrow{BP}$ intersects $\overrightarrow{CD}$. Thus, m intersects $\overrightarrow{CD}$, since $\overrightarrow{BP} \subseteq m$. ∎

Theorem 16: If a line is parallel to an outer side of a trilateral and contains a point in the trilateral's interior, then it intersects the inner side.

Proof: Suppose for trilateral *TABCD* point *P* is in the interior of *TABCD* and lies on a line *m* which is parallel to an outer side of *TABCD*; without loss of generality, suppose *m* is parallel to $\overrightarrow{CD}$.

By Part II material, *B* and *P* determine a line *BP*. Since *B* is a vertex of *TABCD* and *P* ∈ *int*(*TABCD*), then, by Theorem 15, *BP* intersects $\overrightarrow{CD}$ at a point *E*. (See **Figure 11.30**.)

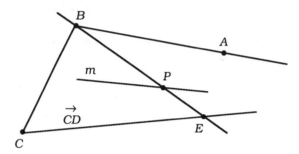

Figure 11.30

Since *P* ∈ *int*(*TABCD*), then, by Part II material (Theorem 21), $P \in \overrightarrow{BE}$ and $P \in \overrightarrow{EB}$. Hence, *P* ∈ (*B*,*E*).

Since *m* is parallel to $\overrightarrow{CD}$, then *m* cannot contain a point of segment (*C*,*E*), since $(C,E) \subseteq \overrightarrow{CD}$.

By Part II material (Axiom 11), *m* contains a point of side (*B*,*C*) of Δ*BCE*. Thus, *m* intersects the inner side of *TABCD*. ■

Theorem 17: If a line intersects a side of a trilateral, does not contain a vertex, and is not parallel to an outer side, then it intersects exactly one of the other two sides.

Proof: Suppose for trilateral *TABCD* line *m* intersects a side, does not contain a vertex, and is not parallel to an outer side.

There are three possibilities:

(a) *m* intersects inner side (*B*,*C*);

(b) *m* intersects outer side $\overrightarrow{BA}$;

(c) *m* intersects outer side $\overrightarrow{CD}$.

Claim: If m intersects inner side (B,C), then m intersects exactly one of $\overrightarrow{BA}$ and $\overrightarrow{CD}$.

Proof: Suppose m intersects side (B,C) at a point E.

By Axiom 1 and Definition 1, there exists a halfline $\overrightarrow{EF}$ which lies on the same side of BC as $\overrightarrow{BA}$ and is parallel to $\overrightarrow{BA}$. (See **Figure 11.31**.) By Theorem 11, $\overrightarrow{EF}$ and $\overrightarrow{CD}$ are parallel to each other.

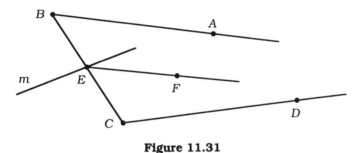

Figure 11.31

By Definition 3, the union of $\{B,E\}$, (B,E), $\overrightarrow{BA}$, and $\overrightarrow{EF}$ is trilateral **T**$ABEF$. Similarly, the union of $\{E,C\}$, (E,C), $\overrightarrow{EF}$, and $\overrightarrow{CD}$ is trilateral **T**$FECD$.

Since m is not parallel to $\overrightarrow{BA}$, then $\overrightarrow{EF}$ does not lie on m. Thus, E is the only point which lies on both m and line EF.

Since $E \in (B,C)$, $E \in EF$, $B \notin EF$, and $C \notin EF$, then, by Part II material (Theorem 20), B and C are on opposite sides of EF.

By Part II material (Theorem 12 and Theorem 19), there exists a point G which lies on m and on the side of BC which contains A and D. Since G is on m and distinct from E, then, by Part II material (Theorem 19), G lies on the B-side of EF or G lies on the C-side of EF, but not both.

Subclaim: If G lies on the B-side of EF, then m intersects $\overrightarrow{BA}$.

Proof: Suppose G lies on the B-side of EF.
By Definition 1, F is on the A-side of BC. Thus, by Part II material (Theorem 22), G is on the side of BC which contains A and F.

Since G lies on the B-side of EF and on the F-side of BC, then $G \in int(\angle BEF)$. Thus, by Part II material (Theorem 40 and Theorem 41), $\overrightarrow{EG}$ intersects segment (B,F) at a point H. (See **Figure 11.32**.) Since $H \in \overrightarrow{EG}$ and $H \in (B,F)$, then $H \in int(\angle BEF)$ and $H \in \overrightarrow{BF}$.

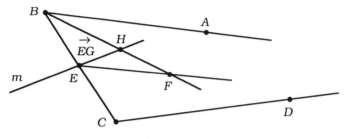

Figure 11.32

Thus, H is on the F-side of BC, the B-side of EF and, by Part II material (Theorem 21), the F-side of BA. Hence, by Definition 4, $H \in int(\mathbf{T}ABEF)$. Since $H \in \overrightarrow{EG}$ and $\overrightarrow{EG} \subseteq m$, then $H \in m$.

Therefore, for **T**ABEF, m passes through vertex E and contains point H in the interior of **T**ABEF. By Theorem 15, m intersects $\overrightarrow{BA}$. ♦♦

Subclaim: If G lies on the C-side of EF, then m intersects $\overrightarrow{CD}$.

Proof: Analogous to the preceding **Subclaim**. ♦♦

Since G lies on the B-side of EF or G lies on the C-side of EF, but not both, then m intersects exactly one of $\overrightarrow{BA}$ and $\overrightarrow{CD}$. ♦

Claim: If m intersects outer side $\overrightarrow{BA}$, then m intersects exactly one of (B,C) and $\overrightarrow{CD}$.

Proof: Suppose m intersects outer side $\overrightarrow{BA}$ at a point E.
Points E and C lie on a line EC which is distinct from m since $C \notin m$. By Part II material (Theorem 12 and Theorem 19), m contains a point F

which lies on the side of BE which contains C and D.

By Theorem 1, $\overrightarrow{CE}$ is between $\overrightarrow{CB}$ and $\overrightarrow{CD}$. Thus, by Part II material (Theorem 41 and Theorem 22), B and D are on opposite sides of EC.

By Part II material (Theorem 19), F lies on the B-side of EC or F lies on the D-side of EC, but not both.

Subclaim: If F lies on the B-side of EC, then m intersects (B,C).

Proof: Left as an exercise. ♦♦

Subclaim: If F lies on the D-side of EC, then m intersects $\overrightarrow{CD}$.

Proof: Left as an exercise. ♦♦

Since F lies on the B-side of EC or F lies on the D-side of EC, but not both, then m intersects exactly one of (B,C) and $\overrightarrow{CD}$. ♦

Claim: If m intersects outer side $\overrightarrow{CD}$, then m intersects exactly one of (B,C) and $\overrightarrow{BA}$.

Proof: Analogous to the preceding **Claim**. ♦

Therefore, if a line intersects a side of a trilateral, does not contain a vertex, and is not parallel to an outer side, it intersects exactly one of the two other sides. ■

Language Rule 3: An angle of a trilateral is sometimes called an **interior angle** of the trilateral.

Definition 7: An interior angle of a trilateral which is not adjacent to an exterior angle of the trilateral is called a **remote interior angle** of the exterior angle.

Definition 8: An exterior angle of a trilateral which is not adjacent to an interior angle of the trilateral is called a **remote exterior angle** of the interior angle.

Theorem 18: An interior angle of a trilateral is less than each of its remote exterior angles.

Proof: Suppose $TABCD$ is a trilateral.

By Definition 3, $TABCD$ is the union of $\{B,C\}$, (B,C), $\overrightarrow{BA}$, and $\overrightarrow{CD}$, where B, A are on a halfline $\overrightarrow{XA}$ with $X \bullet B \bullet A$ and C, D are on a halfline $\overrightarrow{YD}$ with $Y \bullet C \bullet D$, and $\overrightarrow{XA}$, $\overrightarrow{YD}$ are parallel. By Part II material (Axiom 5, Axiom 6, and Axiom 1b), there exist points E and F on line BC such that $E \bullet B \bullet C$ and $B \bullet C \bullet F$. (See **Figure 11.33**.)

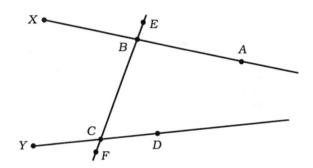

Figure 11.33

By Theorem 12, $\overrightarrow{BA}$ and $\overrightarrow{CD}$ are parallel. By Definition 1, $\overrightarrow{BA}$ is parallel to CD and $\overrightarrow{CD}$ is parallel to BA.

By Definition 3 and Definition 8, each of $\angle YCB$ and $\angle DCF$ is a remote exterior angle for the interior angle $\angle ABC$. Similarly, each of $\angle XBC$ and $\angle ABE$ is a remote exterior angle for the interior angle $\angle DCB$.

Claim: $\angle ABC < \angle YCB$.

Proof: Since $\overrightarrow{XA}$ and $\overrightarrow{YD}$ are parallel, then, by Theorem 8, $\angle ABC \not\equiv \angle YCB$.

Thus, by Part II material (namely, Theorem 64), exactly one of $\angle ABC < \angle YCB$ and $\angle YCB < \angle ABC$ is true.

Subclaim: $\angle YCB < \angle ABC$ is not true.

Proof: Suppose $\angle YCB < \angle ABC$.

By Part II material, there exists a halfline $\overrightarrow{BZ}$ such that $\overrightarrow{BZ}$ is between $\overrightarrow{BA}$ and $\overrightarrow{BC}$ and $\angle CBZ \cong \angle YCB$.

Since $\overrightarrow{BA}$ is parallel to CD, $\overrightarrow{BC}$ intersects CD at C, and $\overrightarrow{BZ}$ is between $\overrightarrow{BA}$ and $\overrightarrow{BC}$, then, by Theorem 1, $\overrightarrow{BZ}$ intersects CD at a point G. By Part II material (Theorem 21 and Theorem 22), $G \in \overrightarrow{CD}$. (See **Figure 11.34**.)

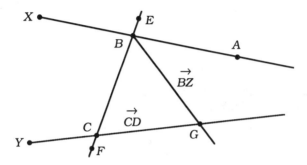

Figure 11.34

Hence, for $\triangle BCG$, $\angle YCB$ is a remote exterior angle for interior angle $\angle CBG$. Since $\overrightarrow{BZ} = \overrightarrow{BG}$, then $\angle CBG \cong \angle YCB$. Thus, Theorem 84 of Part II is contradicted.

Thus, $\angle YCB \not< \angle ABC$. ♦♦

Therefore, $\angle ABC < \angle YCB$. ♦

Claim: $\angle DCB < \angle XBC$.

Proof: Analogous to the preceding **Claim**. ♦

Claim: $\angle ABC < \angle DCF$.

Proof: Left as an exercise. ♦

Claim: $\angle DCB < \angle ABE$.

Proof: Analogous to the preceding **Claim**. ♦

Therefore, an interior angle of a trilateral is less than each of its remote exterior angles. ∎

Definition 9: If $TABCD$ and $TEFGH$ are trilaterals such a one-to-one correspondence can be established between their vertices such that the inner sides are congruent and corresponding angles are congruent, then $TABCD$ and $TEFGH$ are called **congruent trilaterals**.

Notation Rule 3: If $TABCD \leftrightarrow TEFGH$, then the following correspondences occur among their respective parts:

$$B \leftrightarrow F \qquad \angle B \leftrightarrow \angle F \qquad \vec{BA} \leftrightarrow \vec{FE}$$

$$C \leftrightarrow G \qquad \angle C \leftrightarrow \angle G \qquad \vec{CD} \leftrightarrow \vec{GH}$$

$$(B,C) \leftrightarrow (F,G)$$

Theorem 19: If an angle and the inner side of one trilateral are congruent to an angle and the inner side of a second trilateral, then the trilaterals are congruent.

Proof: Suppose for trilaterals $TABCD$ and $TEFGH$, $\angle ABC \cong \angle EFG$ and $(B,C) \cong (F,G)$.

By Part II material (Theorem 64), exactly one of $\angle BCD < \angle FGH$, $\angle BCD \cong \angle FGH$, and $\angle FGH < \angle BCD$ is true.

Claim: $\angle BCD < \angle FGH$ is not true.

Proof: Left as an exercise. ♦

Claim: $\angle FGH < \angle BCD$ is not true.

Proof: Analogous to the preceding **Claim**. ♦

Since $\angle BCD \not< \angle FGH$ and $\angle FGH \not< \angle BCD$, then $\angle BCD \cong \angle FGH$. Therefore, by Definition 9, $TABCD \cong TEFGH$. ∎

Theorem 20: If the angles of one trilateral are congruent, respectively, to the angles of a second trilateral, then the trilaterals are congruent.

Proof: Suppose for trilaterals $TABCD$ and $TEFGH$, $\angle ABC \cong \angle EFG$ and $\angle BCD \cong \angle FGH$.

By Part II material (namely, Theorem 60), exactly one of $(B,C) < (F,G)$, $(B,C) \cong (F,G)$, and $(F,G) < (B,C)$ is true.

Claim: $(B,C) < (F,G)$ is not true.

Proof: Suppose $(B,C) < (F,G)$.

By Part II material, there exists a point K such that $F \bullet K \bullet G$ and $(B,C) \cong (F,K)$. Since $K \in (F,G)$, then $K \in \vec{FG}$. Hence, $\vec{FK} = \vec{FG}$.

By Axiom 1, there exists a halfline $\vec{KM}$ such that $\vec{KM}$ is parallel to $\vec{FE}$. By Definition 1, $\vec{KM}$ is parallel to $\vec{FE}$. Hence, by Theorem 11, $\vec{KM}$ is parallel to $\vec{GH}$. (See **Figure 11.35**.)

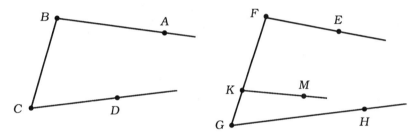

Figure 11.35

By Definition 3, the union of $\{F,K\}$, (F,K), $\vec{FE}$, and $\vec{KM}$ is a trilateral.

In **T**$ABCD$ and **T**$EFKM$, $\angle ABC \cong \angle EFK$ and $(B,C) \cong (F,K)$. Thus, by Theorem 19, **T**$ABCD \cong$ **T**$EFKM$. By Definition 9, $\angle BCD \cong \angle FKM$.

Since $\angle BCD \cong \angle FGH$ and $\angle BCD \cong \angle FKM$, then, by Part II material (Axiom 16), $\angle FGH \cong \angle FKM$. By Theorem 9, $\vec{KM}$ and $\vec{GH}$ are not parallel. Hence, a contradiction, since $\vec{KM}$ and $\vec{GH}$ are both parallel and not parallel.

Therefore, $(B,C) \not< (F,G)$. ◆

Claim: $(F,G) < (B,C)$ is not true.

Proof: Analogous to the preceding **Claim**. ◆

Since $(B,C) \not< (F,G)$ and $(F,G) \not< (B,C)$, then $(B,C) \cong (F,G)$.
Therefore, by Definition 9, **T**ABCD $\cong$ **T**EFGH. ∎

Definition 10: If the angles of a trilateral are congruent, then it is called an **isosceles trilateral**.

Theorem 21: If the inner sides of two isosceles trilaterals are congruent, then the trilaterals are congruent.

Proof: Suppose for two isosceles trilaterals **T**ABCD and **T**EFGH $(B,C) \cong (F,G)$.

Since each of **T**ABCD and **T**EFGH is isosceles, then $\angle ABC \cong \angle BCD$ and $\angle EFG \cong \angle FGH$.

By Part II material (Theorem 64), $\angle ABC < \angle EFG$, $\angle ABC \cong \angle EFG$, or $\angle EFG < \angle ABC$.

Claim: $\angle ABC < \angle EFG$ is not true.

Proof: Suppose $\angle ABC < \angle EFG$.
By Part II material (Theorem 61 and Theorem 62), $\angle BCD < \angle FGH$.

By Part II material there exists a halfline $\overrightarrow{FK}$ such that $\overrightarrow{FK}$ is between $\overrightarrow{FE}$ and $\overrightarrow{FG}$ and $\angle KFG \cong \angle ABC$. Similarly, there exists a halfline $\overrightarrow{GM}$ such that $\overrightarrow{GM}$ is between $\overrightarrow{GF}$ and $\overrightarrow{GH}$ and $\angle FGM \cong \angle BCD$. (See **Figure 11.36**.)

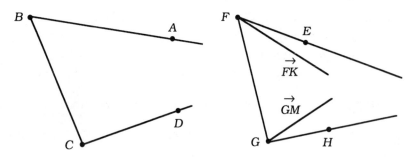

Figure 11.36

By Theorem 12, $\overrightarrow{GH}$ and $\overrightarrow{FE}$ are parallel. By Definition 1, $\overrightarrow{FE}$ is parallel to GH.

Since $\vec{FE}$ is parallel to GH, $\vec{FG}$ intersects GH at G, and $\vec{FK}$ is between $\vec{FE}$ and $\vec{FG}$, then, by Theorem 1, $\vec{FK}$ intersects GH at a point P. By Part II material (Theorem 21 and Theorem 22), $P \in \vec{GH}$. Since $P \in \vec{FK}$ and $P \in \vec{GH}$, then $\vec{FK} = \vec{FP}$ and $\vec{GH} = \vec{GP}$.

Since $\vec{GM} \subseteq int(\angle FGH)$, then, by Part II material (Theorem 41), $\vec{GM}$ intersects (F,P) at a point Q. Since $Q \in \vec{GM}$ and $Q \in (F,P)$, then $\vec{GM} = \vec{GQ}$, $\vec{FP} = \vec{FQ}$, and $\vec{FK} = \vec{FQ}$.

By Part II material (Axiom 12), there exists on $\vec{BA}$ a point R such that $(B,R) \cong (F,Q)$. (See **Figure 11.37**.) Since $R \in \vec{BA}$, then $\vec{BA} = \vec{BR}$.

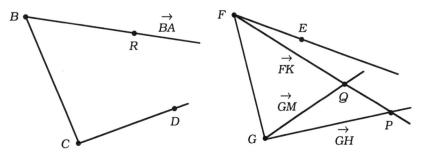

Figure 11.37

In $\triangle CBR$ and $\triangle GFQ$, $(B,C) \cong (F,G)$, $\angle CBR \cong \angle GFQ$, and $(B,R) \cong (F,Q)$. Thus, by Part II material (Theorem 50), $\triangle CBR \cong \triangle GFQ$. Hence, $\angle BCR \cong \angle FGQ$.

Since $\angle BCD \cong \angle FGQ$ and $\angle BCR \cong \angle FGQ$, then, by Part II material (Axiom 16), $\angle BCD \cong \angle BCR$. By Part II material (Theorem 21, Theorem 22, and Axiom 15), $\vec{CR} = \vec{CD}$.

Thus, $\vec{CD}$ is parallel to $\vec{BA}$ and $\vec{CD}$ intersects $\vec{BA}$ at point R. Hence, a contradiction.

Therefore, $\angle ABC \not\cong \angle EFG$. ♦

Claim: $\angle EFG < \angle ABC$ is not true.

Proof: Analogous to the preceding **Claim**. ♦

Since ∠ABC ≮ ∠EFG and ∠EFG ≮ ∠ABC, then ∠ABC ≅ ∠EFG.
By Part II material (Axiom 16), ∠BCD ≅ ∠FGH, since ∠ABC ≅ ∠BCD and ∠EFG ≅ ∠FGH.
Therefore, by Definition 9, **T**ABCD ≅ **T**EFGH. ∎

Theorem 22: The perpendicular bisector of the inner side of a trilateral is parallel to each of the outer sides if and only if the trilateral is isosceles.

Proof: Suppose for trilateral **T**ABCD line m is the perpendicular bisector of inner side (B,C).

Since m is the perpendicular bisector of (B,C), then m intersects (B,C) at a point E such that (B,E) ≅ (E,C). By Part II material (Axiom 2, Axiom 5, Axiom 6, and Axiom 1b), there exist two points F and G such that F•E•G and F, G are on m; without loss of generality, assume $\overrightarrow{EG}$ is on the same side of BC as $\overrightarrow{BA}$ and $\overrightarrow{CD}$. (See **Figure 11.38**.)

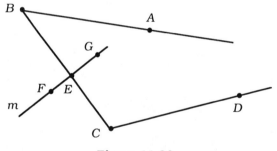

Figure 11.38

Since m is perpendicular to (B,C), then, by Part II material (Theorem 74 and Theorem 70), ∠BEF, ∠BEG, ∠FEC, and ∠GEC are right angles and congruent.

Claim: If m is parallel to each of $\overrightarrow{BA}$ and $\overrightarrow{CD}$, then **T**ABCD is isosceles.

Proof: Left as an exercise. ♦

Claim: If $TABCD$ is isosceles, then m is parallel to each of $\overrightarrow{BA}$ and $\overrightarrow{CD}$.

Proof: Suppose $TABCD$ is isosceles.
By Definition 10, $\angle ABC \cong \angle DCB$.

Since $E \in (B,C)$, then $\overrightarrow{BC} = \overrightarrow{BE}$ and $\overrightarrow{CB} = \overrightarrow{CE}$. Hence, $\angle ABE \cong \angle DCE$. Since E is not on BA, then, by Axiom 1 and Definition 1, there exists a halfline $\overrightarrow{EH}$ which is parallel to $\overrightarrow{BA}$. (See **Figure 11.39**.) By Theorem 10, $\overrightarrow{BA}$ is parallel to $\overrightarrow{EH}$.

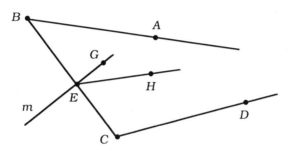

Figure 11.39

By Theorem 12, $\overrightarrow{BA}$ and $\overrightarrow{CD}$ are parallel. Hence, by Theorem 11, $\overrightarrow{CD}$ and $\overrightarrow{EH}$ are parallel.

By Definition 3, the union of $\{B,E\}$, (B,E), $\overrightarrow{BA}$, and $\overrightarrow{EH}$ is a trilateral. Similarly, the union of $\{C,E\}$, (C,E), $\overrightarrow{CD}$, and $\overrightarrow{EH}$ is a trilateral.

In $TABEH$ and $TDCEH$, $\angle ABE \cong \angle DCE$ and $(B,E) \cong (C,E)$. By Theorem 19, $TABEH \cong TDCEH$. Thus, by Definition 9, $\angle BEH \cong \angle CEH$.

Since $\angle BEH$ and $\angle CEH$ are supplementary angles, then each of $\angle BEH$ and $\angle CEH$ is a right angle. Thus, by Part II material (Theorem 70), $\angle BEH \cong \angle BEG$. By Part II material (Theorem 21, Theorem 22, and Axiom 16), $\overrightarrow{EH} = \overrightarrow{EG}$.

Hence, $\overrightarrow{EG}$ is parallel to each of $\overrightarrow{BA}$ and $\overrightarrow{CD}$. Thus, by Definition 1, m is parallel to each of $\overrightarrow{BA}$ and $\overrightarrow{CD}$, since $\overrightarrow{EG} \subseteq m$. ♦

Therefore, the perpendicular bisector of the inner side of a trilateral is parallel to each of the outer sides if and only if the trilateral is isosceles. ∎

Exercises 9.3

1. Develop a proof for Theorem 12.

2. Develop a proof for Theorem 13.

3. Develop a proof for Theorem 14.

4. Complete the proof for Theorem 17.

5. Complete the proof for Theorem 18.

6. Complete the proof for Theorem 19.

7. Complete the proof for Theorem 22.

11.4 Quadrilaterals

A study of plane hyperbolic geometry would be incomplete without an examination of two types of quadrilaterals. Both types were developed and studied in association with efforts to establish that Euclid's parallel axiom is actually a theorem. They are named in honor of their developers, the Italian mathematician Girolamo Saccheri (1667-1733) and the German mathematician Johann Heinrich Lambert (1728-1777).

Definition 11: If A, B, C, D are four points, no three of which are collinear, and no pair of the resulting segments (A,B), (B,C), (C,D), (D,A) intersect, then the set which is the union of $\{A,B,C,D\}$, (A,B), (B,C), (C,D), and (D,A) is called a **quadrilateral**. The points A, B, C, D are called **vertices**, the segments (A,B), (B,C), (C,D), (D,A) are called **sides**, and the lines AB, BC, CD, DA are called **sidelines**.

It should be observed that the vertices of a quadrilateral should be named by beginning at any vertex and proceeding around the quadrilateral in one fixed direction, either clockwise or counterclockwise.

Notation Rule 4: If A, B, C, D are the vertices of a quadrilateral then the quadrilateral may be denoted as $QABCD$, $QBCDA$, $QCDAB$, or $QDABC$.

Definition 12: In a quadrilateral, two vertices which are endpoints of the same side are called **consecutive vertices**, while two vertices which are not consecutive are called **opposite vertices**. A segment which has a pair of opposite vertices as its endpoints is called a **diagonal**. Two sides which have a common endpoint are called **adjacent sides**, while two sides which are not adjacent are called **opposite sides**.

Language Rule 4: In a quadrilateral, **angle of the quadrilateral** will refer to an angle which has for its vertex a vertex of the quadrilateral and whose sides contain the sides of the quadrilateral for which the vertex is the common endpoint. **Consecutive angles** will refer to two angles whose intersection contains a side of the quadrilateral, while two angles which are not consecutive will be referred to as **opposite angles**.

Definition 13: If for each sideline of a quadrilateral the vertices not on that sideline lie on the same side of the sideline, then the quadrilateral is called a **convex quadrilateral**.

Definition 14: A convex quadrilateral $QABCD$ in which $\angle A$ and $\angle B$ are right angles and $(A,D) \cong (B,C)$ is called a **Saccheri quadrilateral**. Segment (A,B) is called the **base**, (D,C) is called the **summit**, and $\angle C$ and $\angle D$ are called the **summit angles**.

Notation Rule 5: A convex quadrilateral $QABCD$ which is a Saccheri quadrilateral may be denoted as $SQABCD$. In the symbol $SQABCD$, points A and B are the vertices of the right angles.

Theorem 23: The line containing the midpoints of the base and summit of a Saccheri quadrilateral is perpendicular to each of them.

Theorem 24: The base and summit sidelines of a Saccheri quadrilateral are not parallel and do not intersect.

Proof: Suppose $SQABCD$ is a Saccheri quadrilateral.

By Part II material (Theorem 88), (A,B) and (D,C) have midpoints E and F, respectively. By Theorem 23, line EF is perpendicular to each of (A,B) and (D,C). (See **Figure 11.40**.)

By Part II material (Theorem 74 and Theorem 70), $\angle AEF$, $\angle BEF$, $\angle CFE$ are right angles and congruent.

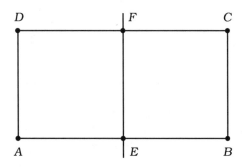

Figure 11.40

Claim: Lines AB and DC are not parallel.

Proof: Since $\angle AEF$ and $\angle CFE$ are congruent alternate interior angles formed when halflines $\overrightarrow{AB}$ and $\overrightarrow{DC}$ are intersected by transversal EF, then $\overrightarrow{AB}$ and $\overrightarrow{DC}$ are not parallel, by Theorem 8. By Definition 1 and Axiom 1, lines AB and DC are not parallel. ◆

Claim: Lines AB and DC do not intersect.

Proof: Suppose AB and DC intersect at a point G.

By Definition 11, $G \notin (A,B)$, $G \notin (D,C)$, $G \notin \{A,B,C,D\}$. Thus, $A \bullet B \bullet G$, $D \bullet C \bullet G$ or $G \bullet A \bullet B$, $G \bullet D \bullet C$; without loss of generality, assume $A \bullet B \bullet G$ and $D \bullet C \bullet G$ (See **Figure 11.41**.)

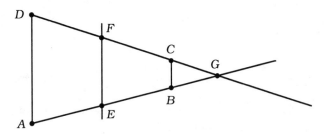

Figure 11.41

Since $A \bullet E \bullet B$ and $A \bullet B \bullet G$, then, by Part II material (Theorem 7 and Axiom 9), $E \bullet B \bullet G$. Similarly, $F \bullet C \bullet G$. Hence, $\overrightarrow{EB} = \overrightarrow{EG}$ and $\overrightarrow{FC} = \overrightarrow{FG}$. Thus, $\angle BEF = \angle GEF$ and $\angle CFE = \angle GFE$.

In $\triangle FEG$, each of $\angle GEF$ and $\angle GFE$ is a right angle. This contradicts

Theorem 96 of Part II.
  Hence, AB and DC do not intersect. ◆

Therefore, the base and summit sidelines of a Saccheri quadrilateral are not parallel and do not intersect. ∎

Theorem 25: The summit angles of a Saccheri quadrilateral are congruent.

Proof: Suppose $S\!Q\!ABCD$ is a Saccheri quadrilateral.
By Definition 14, $(A,D) \cong (B,C)$.
By Part II material (Theorem 88), there exists exactly one point E such that $E \in (A,B)$ and $(A,E) \cong (B,E)$. Similarly, there exists exactly one point F such that $F \in (D,C)$ and $(D,F) \cong (C,F)$. (see **Figure 11.42**.) Since $F \in (D,C)$, then $F \in \overrightarrow{DC}$ and $F \in \overrightarrow{CD}$. Thus, $\overrightarrow{DF} = \overrightarrow{DC}$ and $\overrightarrow{CF} = \overrightarrow{CD}$.

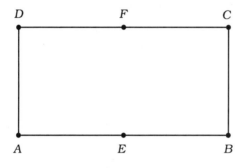

Figure 11.42

By Theorem 23, line EF is perpendicular to (A,B). Hence, by Part II material (Theorem 74 and Theorem 70), $\angle AEF$ and $\angle BEF$ are right angles and congruent.

In $\triangle AEF$ and $\triangle BEF$, $(A,E) \cong (B,E)$, $\angle AEF \cong \angle BEF$, and $(E,F) \cong (E,F)$. Thus, by Part II material (Theorem 50), $\triangle AEF \cong \triangle BEF$. Therefore, $(A,F) \cong (B,F)$.

In $\triangle ADF$ and $\triangle BCF$, $(A,D) \cong (B,C)$, $(D,F) \cong (C,F)$, and $(A,F) \cong (B,F)$. Thus, by Part II material (Theorem 56), $\triangle ADF \cong \triangle BCF$. Therefore, $\angle ADF \cong \angle BCF$.

Since $\angle ADF = \angle ADC$ and $\angle BCF = \angle BCD$, then $\angle ADC \cong \angle BCD$. Therefore, the summit angles of $S\!Q\!ABCD$ are congruent. ∎

Theorem 26: The summit angles of a Saccheri quadrilateral are acute.

Proof: Suppose $SQABCD$ is a Saccheri quadrilateral.
By Definition 14, $(A,D) \cong (B,C)$, and $\angle DAB$, $\angle CBA$ are right angles.

By Part II material (Axiom 5, Axiom 6, and Axiom 1b), there exist points E and F such that $A \bullet B \bullet E$, $E \in AB$, $D \bullet C \bullet F$, and $F \in CD$. By Part II material (Theorem 74 and Theorem 70), $\angle CBE$ is a right angle and $\angle CBE$, $\angle DAB$, $\angle CBA$ are congruent.

By Axiom 1 and Definition 1, there exist halflines $\vec{CG}$ and $\vec{DH}$ such that $\vec{CG}$ is parallel to $\vec{BE}$ from C and $\vec{DH}$ is parallel to $\vec{AB}$ from D. By Theorem 10, $\vec{BE}$ is parallel to $\vec{CG}$.

Since $A \bullet B \bullet E$ and $\vec{BE}$ is parallel to $\vec{CG}$, then $\vec{AB}$ and $\vec{CG}$ are parallel, by Theorem 7 and Theorem 10. By Theorem 11, $\vec{CG}$ is parallel to $\vec{DH}$, since each of $\vec{DH}$ and $\vec{CG}$ is parallel to $\vec{AB}$.

By Theorem 24 and Definition 1, $\vec{CF} \neq \vec{CG}$ and $\vec{DC} \neq \vec{DH}$. Also, by Theorem 24, $\vec{CF}$ and $\vec{DC}$ do not intersect AB. Thus, by Theorem 1, $\vec{CF}$ is not between $\vec{CG}$ and $\vec{CB}$, and $\vec{DC}$ is not between $\vec{DH}$ and $\vec{DA}$. (See **Figure 11.43**.)

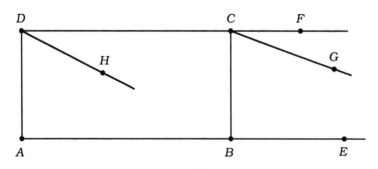

Figure 11.43

By Definition 3, the union of $\{A,D\}$, (A,D), $\vec{AB}$, and $\vec{DH}$ is a trilateral.

Similarly, the union of $\{B,C\}$, (B,C), $\overrightarrow{BE}$, and $\overrightarrow{CG}$ is a trilateral. Also, the union of $\{C,D\}$, (C,D), $\overrightarrow{CG}$, and $\overrightarrow{DH}$ is a trilateral.

For **T**HDAB and **T**GCBE, $(A,D) \cong (B,C)$ and $\angle DAB \cong \angle CBE$. Thus, by Theorem 19, **T**HDAB $\cong$ **T**GCBE. Thus, by Definition 9, $\angle HDA \cong \angle GCB$.

For **T**GCDH, $\angle CDH$ is a remote interior angle for exterior angle $\angle FCG$. By Theorem 18, $\angle CDH < \angle FCG$. By Part II material, there exists a halfline $\overrightarrow{CK}$ such that $\overrightarrow{CK}$ is between $\overrightarrow{CF}$ and $\overrightarrow{CG}$ and $\angle CDH \cong \angle KCG$.

Since $\angle CDH \cong \angle KCG$ and $\angle HDA \cong \angle GCB$, then, by Part II material (Theorem 53), $\angle CDA \cong \angle KCB$.

Claim: $\angle KCB < \angle FCB$.

Proof: Left as an exercise. ♦

Hence, by Part II material (Theorem 61), $\angle CDA < \angle FCB$, since $\angle CDA \cong \angle KCB$ and $\angle KCB < \angle FCB$.

Since $\angle CDA \cong \angle DCB$, by Theorem 25, and $\angle CDA < \angle FCB$, then, by Part II material (Theorem 61), $\angle DCB < \angle FCB$. Thus, by Part II material. (Theorem 64), $\angle DCB \not\cong \angle FCB$.

By Part II material (Theorem 75, Theorem 73, and Theorem 72), $\angle DCB$ is an acute angle.

Since $\angle DCB$ is acute and $\angle DAB$ is a right angle, then $\angle DCB < \angle DAB$. Thus, by Part II material (Theorem 61), $\angle CDA < \angle DAB$, since $\angle CDA \cong \angle DCB$. Hence, $\angle CDA$ is an acute angle.

Thus, the summit angles of a Saccheri quadrilateral are acute. ∎

Theorem 27: The base of a Saccheri quadrilateral is less than its summit.

Proof: Suppose **SQ**ABCD is a Saccheri quadrilateral.

By Definition 14, $\angle A$ is a right angle. By Theorem 26, $\angle D$ is an acute angle.

By Part II material (Theorem 88), (A,B) and (D,C) have midpoints E and F, respectively. Thus, $A \bullet E \bullet B$, $(A,E) \cong (E,B)$, $D \bullet F \bullet C$, and $(D,F) \cong (F,C)$. By Theorem 23, line EF is perpendicular to each of (A,B) and (D,C). (See **Figure 11.44**.)

By Part II material (Theorem 70), $\angle DAE$, $\angle AEF$, and $\angle DFE$ are congruent.

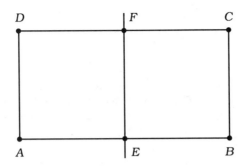

Figure 11.44

By Part II material (Theorem 60), exactly one of $(A,E) < (D,F)$, $(A,E) \cong (D,F)$, and $(D,F) < (A,E)$ is true.

Claim: $(A,E) \cong (D,F)$ is not true.

Proof: Left as an exercise. ♦

Claim: $(D,F) < (A,E)$ is not true.

Proof: Suppose $(D,F) < (A,E)$.

By Part II material, there exists a point G such that $A \bullet G \bullet E$ and $(G,E) \cong (D,F)$. (See **Figure 11.45**.) Since sidelines AB and DC do not intersect, by Theorem 24, $D \neq G$. Since $G \in (A,E)$, then $\overrightarrow{AG} = \overrightarrow{AE}$ and $\overrightarrow{EG} = \overrightarrow{EA}$. Thus, $\angle DAE = \angle DAG$ and $\angle AEF = \angle GEF$.

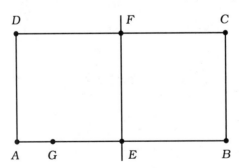

Figure 11.45

In $\triangle DFE$ and $\triangle GEF$, $(D,F) \cong (G,E)$, $\angle DFE \cong \angle GEF$, and $(F,E) \cong (E,F)$.

298 CHAPTER 11

Thus, by Part II material (Theorem 50), $\triangle DFE \cong \triangle GEF$. Hence, $(D,E) \cong (G,F)$.

In $\triangle GDF$ and $\triangle DGE$, $(G,D) \cong (D,G)$, $(D,F) \cong (G,E)$, and $(G,F) \cong (D,E)$. Thus, by Part II material (Theorem 56), $\triangle GDF \cong \triangle DGE$. Hence, $\angle FDG \cong \angle EGD$.

For $\triangle ADG$, $\angle EGD$ is a remote exterior angle for the interior angle $\angle DAG$. By Part II material (Theorem 84), $\angle DAG < \angle EGD$. Since $\angle FDG \cong \angle EGD$, then, by Part II material (Theorem 62), $\angle DAG < \angle FDG$.

By Part II material (Theorem 64), $\angle DAG \not\cong \angle FDG$ and $\angle FDG \not< \angle DAG$. Thus, $\angle FDG$ is an obtuse angle, since $\angle DAG$ is a right angle.

By Part II material (Theorem 21, Theorem 22, and Theorem 40), $G \in int(\angle FDA)$ and $\overrightarrow{DG} \subseteq int(\angle FDA)$. Hence, $\overrightarrow{DG}$ is between $\overrightarrow{DF}$ and $\overrightarrow{DA}$. Thus, $\angle FDG < \angle FDA$.

Hence, a contradiction of Theorem 72 of Part II, since $\angle FDG$ is obtuse and $\angle FDA$ is acute. Therefore, $(D,F) \not< (A,E)$. ♦

Thus, $(A,E) < (D,F)$.

By Part II material, there exists a point H such that $D \bullet H \bullet F$ and $(A,E) \cong (H,F)$.

Since $(A,E) \cong (E,B)$, $(D,F) \cong (F,C)$, and $(A,E) < (D,F)$, then, by Part II material (Theorem 57 and Theorem 58), $(E,B) < (F,C)$. By Part II material, there exists a point J such that $F \bullet J \bullet C$ and $(E,B) \cong (F,J)$.

Since $D \bullet F \bullet C$ and $F \bullet J \bullet C$, then, by Part II material (Theorem 8 and Axiom 9), $D \bullet F \bullet J$ and $D \bullet J \bullet C$. Since $D \bullet F \bullet J$ and $D \bullet H \bullet F$, then, by Part II material (Theorem 7 and Axiom 9), $H \bullet F \bullet J$ and $D \bullet H \bullet J$.

Since $A \bullet E \bullet B$, $H \bullet F \bullet J$, $(A,E) \cong (H,F)$, and $(E,B) \cong (F,J)$, then, by Part II material (Axiom 14), $(A,B) \cong (H,J)$.

Since $(A,B) \cong (H,J)$ and $D \bullet H \bullet J$, then $(A,B) < (D,J)$. Also, since $D \bullet J \bullet C$, then $(D,J) < (D,C)$. Thus, by Part II material (Theorem 59), $(A,B) < (D,C)$. Therefore, the base of a Saccheri quadrilateral is less than its summit. ∎

Theorem 28: The segment whose endpoints are the midpoints of the base and summit of a Saccheri quadrilateral is less than each of the other two sides.

Theorem 29: The diagonals of a Saccheri quadrilateral are congruent.

Theorem 30: The diagonals of a Saccheri quadrilateral do not bisect each other.

Proof: Suppose **SQ**$ABCD$ is a Saccheri quadrilateral.

By Definition 14, $(A,D) \cong (B,C)$ and $\angle DAB$, $\angle CBA$ are right angles. By Part material (Theorem 70), $\angle DAB$ and $\angle CBA$ are congruent. By Theorem 26, $\angle ADC$, $\angle BCD$ are acute angle. By Theorem 29, $(A,C) \cong (B,D)$.

By Part II material (Theorem 21, Theorem 40, and Theorem 41), $C \in int(\angle DAB)$, $\overrightarrow{AC} \subseteq int(\angle DAB)$, and $\overrightarrow{AC}$ intersects (B,D) at a point E. Analogously, $\overrightarrow{BD}$ intersects (A,C) at a point F. Also, $\overrightarrow{AC}$ is between $\overrightarrow{AD}$, $\overrightarrow{AB}$ and $\overrightarrow{BD}$ is between $\overrightarrow{BC}$, $\overrightarrow{BA}$.

Since $E \in (B,D)$ and $E \in \overrightarrow{AC}$, then $E \in BD$ and $E \in AC$. Since $F \in (A,C)$ and $F \in \overrightarrow{BD}$, then $F \in AC$ and $F \in BD$. By Part II material (Theorem 2), $E = F$. (See **Figure 11.46**.)

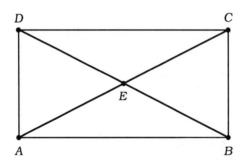

Figure 11.46

In $\triangle ABC$ and $\triangle BAD$, $(A,B) \cong (B,A)$, $\angle CBA \cong \angle DAB$, and $(B,C) \cong (A,D)$. Thus, by Part II material (Theorem 50), $\triangle ABC \cong \triangle BAD$. Hence, $\angle CAB \cong \angle DBA$.

Since $\overrightarrow{AC}$ is between $\overrightarrow{AD}$ and $\overrightarrow{AB}$, $\overrightarrow{BD}$ is between $\overrightarrow{BC}$ and $\overrightarrow{BA}$, $\angle DAB \cong \angle CBA$, and $\angle CAB \cong \angle DBA$, then, by Part II material (Theorem 54), $\angle DAC \cong \angle CBD$. Since $E \in \overrightarrow{AC}$ and $E \in \overrightarrow{BD}$, then $\overrightarrow{AC} = \overrightarrow{AE}$, $\overrightarrow{BD} = \overrightarrow{BE}$, and $\angle DAE \cong \angle CBE$.

By Part II material (Theorem 68), $\angle AED \cong \angle BEC$ and $\angle AEB \cong \angle DEC$, since they are pairs of vertical angles.

In $\triangle AED$ and $\triangle BEC$, $(A,D) \cong (B,C)$, $\angle DAE \cong \angle CBE$, and $\angle AED \cong \angle BEC$. Thus, by Part II material (Theorem 87), $\triangle AED \cong \triangle BEC$. Hence, $(A,E) \cong (B,E)$ and $(E,C) \cong (E,D)$.

Claim: $(A,E) \cong (E,C)$ is not true.

Proof: Left as an exercise. ♦

Claim: $(B,E) \cong (E,D)$ is not true.

Proof: Analogous to the preceding **Claim**. ♦

Therefore, the diagonals of a Saccheri quadrilateral do not bisect each other. ∎

Theorem 31: If a pair of consecutive angles in a convex quadrilateral are right angles and the other pair of angles are congruent, then the quadrilateral is a Saccheri quadrilateral.

Proof: Suppose $\mathcal{Q}ABCD$ is a convex quadrilateral such that $\angle A$ and $\angle B$ are right angles and $\angle C \cong \angle D$.

By Part II material (Theorem 60), exactly one of $(A,D) < (B,C)$, $(B,C) < (A,D)$, and $(A,D) \cong (B,C)$ is true.

Claim: $(A,D) < (B,C)$ is not true.

Proof: Suppose $(A,D) < (B,C)$.

By Part II material, there exists a point E such that $B \bullet E \bullet C$ and $(A,D) \cong (B,E)$. (See **Figure 11.47**.)

In $\mathcal{Q}ABED$, $\angle A$ and $\angle B$ are right angles and $(A,D) \cong (B,E)$. Thus, $\mathcal{Q}ABED$ is a Saccheri quadrilateral, by Definition 14.

In $S\mathcal{Q}ABED$, $\angle ADE \cong \angle BED$, by Theorem 25.

Since $\mathcal{Q}ABCD$ is convex, then, by Part II material (Theorem 21 and Theorem 40), $\overrightarrow{DE}$ is between $\overrightarrow{DA}$ and $\overrightarrow{DC}$. Hence, by Part II material, $\angle ADE < \angle ADC$. Since $\angle ADE \cong \angle BED$ and $\angle ADE < \angle ADC$, then, by Part II material (Theorem 61), $\angle BED < \angle ADC$.

For $\triangle DCE$, $\angle BED$ is a remote exterior angle for the interior angle $\angle DCE$.

Figure 11.47

By Part II material (Theorem 84), $\angle DCE < \angle BED$.

Thus, $\angle DCE < \angle BED$ and $\angle BED < \angle ADC$. Hence, by Part II material (Theorem 63), $\angle DCE < \angle ADC$. Since $E \in (C,B)$, then $E \in \overrightarrow{CB}$ and $\overrightarrow{CE} = \overrightarrow{CB}$. Thus, $\angle DCB < \angle ADC$.

This is a contradiction, since $\angle DCB \cong \angle ADC$ and $\angle DCB < \angle ADC$.
Therefore, $(A,D) \not< (B,C)$. ♦

Claim: $(B,C) < (A,D)$ is not true.

Proof: Analogous to the preceding **Claim**. ♦

Hence, $(A,D) \cong (B,C)$.

Therefore, in convex quadrilateral $\mathcal{Q}ABCD$, $\angle A$ and $\angle B$ are right angles and $(A,D) \cong (B,C)$.

Thus, by Definition 14, $\mathcal{Q}ABCD$ is a Saccheri quadrilateral. ∎

Definition 15: A convex quadrilateral which has three right angles is called a **Lambert quadrilateral**.

Notation Rule 6: A quadrilateral $\mathcal{Q}ABCD$ which is a Lambert quadrilateral may be denoted as $L\mathcal{Q}ABCD$. In the symbol $L\mathcal{Q}ABCD$, points A, B, C are the vertices of the right angles.

Theorem 32: The fourth angle of a Lambert quadrilateral is an acute angle.

Proof: Suppose $L\mathcal{Q}ABCD$ is a Lambert quadrilateral.
By Definition 15, $\angle A$, $\angle B$, $\angle C$ are right angles.

By Part II material (Axiom 12), there exist on sidelines AB and DC points E and F, respectively, such that $A \bullet B \bullet E$, $D \bullet C \bullet F$, $(A,B) \cong (B,E)$, and $(D,C) \cong (C,F)$. (See **Figure 11.48**.) By Part II material (Theorem 74). $\angle BCF$ and $\angle CBE$ are right angles.

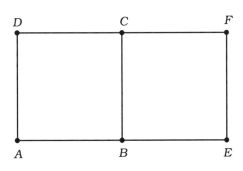

Figure 11.48

By Part II material (Theorem 70), $\angle DAB$, $\angle ABC$, $\angle BCD$, $\angle BCF$, and $\angle EBC$ are congruent.

By Part II material (Theorem 21 and Theorem 40), $\overrightarrow{BD}$ is between $\overrightarrow{BA}$ and $\overrightarrow{BC}$. Similarly, $\overrightarrow{BF}$ is between $\overrightarrow{BC}$ and $\overrightarrow{BE}$.

Claim: Halflines $\overrightarrow{BE}$ and $\overrightarrow{CF}$ do not intersect.

Proof: Left as an exercise. ◆

In $\triangle BCD$ and $\triangle BCF$, $(D,C) \cong (F,C)$, $\angle BCD \cong \angle BCF$, and $(B,C) \cong (B,C)$. Thus, by Part II material (Theorem 50), $\triangle BCD \cong \triangle BCF$. Hence, $\angle CBD \cong \angle CBF$ and $(B,D) \cong (B,F)$.

Since $\overrightarrow{BD}$ is between $\overrightarrow{BC}$ and $\overrightarrow{BA}$, $\overrightarrow{BF}$ is between $\overrightarrow{BC}$ and $\overrightarrow{BE}$, $\angle CBD \cong \angle CBF$, and $\angle ABC \cong \angle EBC$, then, by Part II material (Theorem 54), $\angle ABD \cong \angle EBF$.

In $\triangle ABD$ and $\triangle EBF$, $(A,B) \cong (E,B)$, $\angle ABD \cong \angle EBF$, and $(B,D) \cong (B,F)$. Thus, by Part II material (Theorem 50), $\triangle ABD \cong \triangle EBF$. Hence, $(A,D) \cong (E,F)$ and $\angle DAB \cong \angle FEB$. By Part II material (Theorem 69), $\angle FEB$ is a right angle.

Thus, in ⃞$AEFD$, $(A,D) \cong (E,F)$ and $\angle DAB$, $\angle FEB$ are right angles.

Hence, $\mathcal{Q}AEFD$ is a Saccheri quadrilateral, by Definition 14.

By Theorem 26, $\angle ADF$ is acute, since it is a summit angle for $\mathbf{SQ}AEFD$.

Since $C \in (D,F)$, then $\overrightarrow{DC} = \overrightarrow{DF}$. Thus, $\angle ADC$ is acute. ∎

Theorem 33: A side of a Lambert quadrilateral which does not lie on a side of the acute angle is less than its opposite side.

Proof: Suppose $\mathbf{LQ}ABCD$ is a Lambert quadrilateral.

By Definition 15, $\angle A$, $\angle B$, $\angle C$ are right angles. By Part II material (Theorem 70), $\angle A$, $\angle B$, $\angle C$ are congruent.

By Theorem 32, $\angle ADC$ is acute. Hence, (A,B) and (B,C) are the sides of $\mathbf{LQ}ABCD$ which do not lie on a side of $\angle ADC$. By Definition 12, side (D,C) is opposite (A,B) and side (D,A) is opposite (C,B).

Claim: $(A,B) < (D,C)$.

Proof: By Part II material (Theorem 60), exactly one of $(A,B) < (D,C)$, $(A,B) \cong (D,C)$, and $(D,C) < (A,B)$ is true.

Subclaim: $(A,B) \cong (D,C)$ is not true.

Proof: Left as an exercise. ◆◆

Subclaim: $(D,C) < (A,B)$ is not true.

Proof: Suppose $(D,C) < (A,B)$.

By Part II material, there exists a point E such that $A \bullet E \bullet B$ and $(E,B) \cong (D,C)$. (See **Figure 11.49**.) Since $\mathcal{Q}ABCD$ is convex, then, by Part II material (Theorem 31), E is on the side of line DC containing A and B. Hence, $E \ne D$.

In quadrilateral $\mathcal{Q}BCDE$, $(E,B) \cong (D,C)$ and $\angle EBC$, $\angle DCB$ are right angles. Hence, it is a Saccheri quadrilateral, by Definition 14.

By Theorem 26, the summit angles $\angle BED$ and $\angle CDE$ are acute angles.

For $\triangle ADE$, $\angle BED$ is a remote exterior angle for interior angle $\angle DAE$. By Part II material (Theorem 84), $\angle DAE < \angle BED$.

Hence, a contradiction of Part II material, since $\angle DAE$ is right and $\angle BED$ is acute. Therefore, $(D,C) \not< (A,B)$. ◆◆

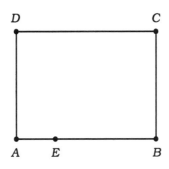

Figure 11.49

Therefore, $(A,B) < (D,C)$. ◆

Claim: $(C,B) < (D,A)$.

Proof: Analogous to the preceding **Claim**. ◆

Thus, a side of a Lambert quadrilateral which does not lie on a side of the acute angle is less than its opposite side. ∎

Definition 16: If $\mathbf{Q}ABCD$ and $\mathbf{Q}EFGH$ are quadrilaterals such that a one-to-one correspondence can be established between their vertices such that corresponding sides are congruent and corresponding angles are congruent, then $\mathbf{Q}ABCD$ and $\mathbf{Q}EFGH$ are called **congruent quadrilaterals**.

Notation Rule 7: If $\mathbf{Q}ABCD \leftrightarrow \mathbf{Q}EFGH$, then the following correspondences occur among their respective parts:

$A \leftrightarrow E \qquad \angle A \leftrightarrow \angle E \qquad (A,B) \leftrightarrow (E,F)$

$B \leftrightarrow F \qquad \angle B \leftrightarrow \angle F \qquad (B,C) \leftrightarrow (F,G)$

$C \leftrightarrow G \qquad \angle C \leftrightarrow \angle G \qquad (C,D) \leftrightarrow (G,H)$

$D \leftrightarrow H \qquad \angle D \leftrightarrow \angle H \qquad (D,A) \leftrightarrow (H,E)$

Theorem 34: If $\mathbf{LQ}ABCD$ and $\mathbf{LQ}EFGH$ are two Lambert quadrilaterals such that $(A,B) \cong (E,F)$ and $(B,C) \cong (F,G)$, then $\mathbf{LQ}ABCD \cong \mathbf{LQ}EFGH$.

Theorem 35: If $\mathbf{LQ}ABCD$ and $\mathbf{LQ}EFGH$ are two Lambert quadrilaterals such that a side of $\mathbf{LQ}ABCD$ which lies on a side of $\angle D$ and its adjacent side which does not lie on a side of $\angle D$ are congruent, respectively, to the corresponding sides of $\mathbf{LQ}EFGH$, then $\mathbf{LQ}ABCD \cong \mathbf{LQ}EFGH$.

Theorem 36: If *LQABCD* and *LQEFGH* are two Lambert quadrilaterals such that ∠D ≅ ∠H and a side of *LQABCD* which lies on a side of ∠D is congruent to the corresponding side of *LQEFGH*, then *LQABCD* ≅ *LQEFGH*.

Proof: Suppose *LQABCD* and *LQEFGH* are two Lambert quadrilaterals such that ∠D ≅ ∠H and a side of *LQABCD* which lies on a side of ∠D is congruent to the corresponding side of *LQEFGH*; without loss of generality, suppose $(A,D) \cong (E,H)$.

By Definition 15, ∠A, ∠B, ∠C, ∠E, ∠F, ∠G are right angles. By Part II material (Theorem 70), ∠A, ∠B, ∠C, ∠E, ∠F, ∠G are congruent.

By Part II material (Theorem 60), exactly one of $(A,B) < (E,F)$, $(E,F) < (A,B)$, and $(A,B) \cong (E,F)$ is true.

Claim: $(A,B) < (E,F)$ is not true.

Proof: Suppose $(A,B) < (E,F)$.

By Part II material, there exists a point J such that $E \bullet J \bullet F$ and $(A,B) \cong (E,J)$. Since $E \bullet J \bullet F$, then $J \in EF$. By Part II material (Theorem 92), there exists exactly one line m which contains J and is perpendicular to EF. By Definition 13 and Part II material (Theorem 31), J is on the side of HG which contains E and F. Thus, $J \notin GH$.

Subclaim: Line m intersects (H,G) at a point K.

Proof: Left as an exercise. ♦♦

By Part II material (Theorem 70 and Theorem 74), ∠EJK and ∠FJK are right angles and ∠EJK ≅ ∠B. (See **Figure 11.50**.)

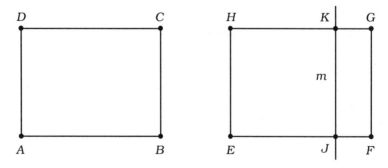

Figure 11.50

By Definition 15, ⌐JFGK is a Lambert quadrilateral.

In $\triangle DAB$ and $\triangle HEJ$, $(A,D) \cong (E,H)$, $\angle A \cong \angle E$, and $(A,B) \cong (E,J)$. Thus, by Part II material (Theorem 50), $\triangle DAB \cong \triangle HEJ$. Hence, $\angle ADB \cong \angle EHJ$, $(D,B) \cong (H,J)$, and $\angle ABD \cong \angle EJH$.

By Part II material (Theorem 21, Theorem 22, and Theorem 40), $\overrightarrow{DB}$ is between $\overrightarrow{DA}$ and $\overrightarrow{DC}$, $\overrightarrow{BD}$ is between $\overrightarrow{BA}$ and $\overrightarrow{BC}$, $\overrightarrow{HJ}$ is between $\overrightarrow{HE}$ and $\overrightarrow{HK}$, and $\overrightarrow{JH}$ is between $\overrightarrow{JE}$ and $\overrightarrow{JK}$.

Since $\angle D \cong \angle H$, $\overrightarrow{DB}$ is between $\overrightarrow{DA}$ and $\overrightarrow{DC}$, $\overrightarrow{HJ}$ is between $\overrightarrow{HE}$ and $\overrightarrow{HK}$, and $\angle ADB \cong \angle EHJ$, then, by Part II material (Theorem 54), $\angle BDC \cong \angle JHK$. Since $\angle B \cong \angle EJK$, $\overrightarrow{BD}$ is between $\overrightarrow{BA}$ and $\overrightarrow{BC}$, $\overrightarrow{JH}$ is between $\overrightarrow{JE}$ and $\overrightarrow{JK}$, and $\angle ABD \cong \angle EJH$, then, by Part II material (Theorem 54), $\angle DBC \cong \angle HJK$.

Thus, $\angle BDC \cong \angle JHK$, $(D,B) \cong (H,J)$, and $\angle DBC \cong \angle HJK$ in $\triangle DBC$ and $\triangle HJK$. Therefore, by Part II material (Theorem 55), $\triangle DBC \cong \triangle HJK$. Hence, $\angle BCD \cong \angle JKH$.

Since $\angle BCD$ is a right angle and $\angle BCD \cong \angle JKH$, then, by Part material (Theorem 69), $\angle JKH$ is a right angle. By Part II material (Theorem 74), $\angle JKG$ is a right angle.

Thus, the fourth angle of L⌐JFGK is not acute. This is a contradiction of Theorem 32.

Hence, $(A,B) \not< (E,F)$. ◆

Claim: $(E,F) < (A,B)$ is not true.

Proof: Analogous to the preceding **Subclaim**. ◆

Therefore, $(A,B) \cong (E,F)$.

Thus, in L⌐ABCD and L⌐EFGH, $(A,D) \cong (E,H)$ and $(A,B) \cong (E,F)$.

Hence, by Theorem 35, L⌐ABCD $\cong$ L⌐EFGH. ■

Theorem 37: If L⌐ABCD and L⌐EFGH are two Lambert quadrilaterals such that $\angle D \cong \angle H$ and a side of L⌐ABCD which does not lie on a side of $\angle D$ is congruent to the corresponding side of L⌐EFGH, then L⌐ABCD $\cong$ L⌐EFGH.

Proof: Suppose $LQABCD$ and $LQEFGH$ are two Lambert quadrilaterals such that $\angle D \cong \angle H$ and a side of $LQABCD$ which does not lie on a side of $\angle D$ is congruent to the corresponding side of $LQEFGH$; without loss of generality, suppose $(A,B) \cong (E,F)$.

By Definition 15, $\angle A$, $\angle B$, $\angle C$, $\angle E$, $\angle F$, $\angle G$ are right angles.

By Part II material (Axiom 12), there exists a point J such that $D \bullet A \bullet J$ and $(A,J) \cong (E,H)$. Since $D \bullet A \bullet J$, then D and J are on opposite sides of AB; thus, $J \neq B$. By Part II material (Theorem 21 and Theorem 22), $J \notin \overrightarrow{BC}$.

If J is on the halfline opposite to $\overrightarrow{BC}$, then $\triangle AJB$ would have two right angles, contradicting Theorem 96 of Part II. Hence, $J \notin BC$.

Since $J \notin BC$, then, by Part II material (Theorem 90), there exists a line m which contains J and is perpendicular to BC at a point K. If $K = B$, then the two lines AB and JB would be perpendicular to BC at B, contradicting Theorem 92 of Part II. If $K \in \overrightarrow{BC}$, then m would intersect (A,B) at a point Z and $\triangle BKZ$ would have two right angles, contradicting Theorem 96 of Part II. Hence, $C \bullet B \bullet K$. (See **Figure 11.51**.)

Angles $\angle JAB$ and $\angle ABK$ are right angles, by Part II material (Theorem 74).

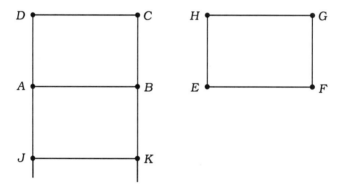

Figure 11.51

Quadrilateral $QKBAJ$ is a Lambert quadrilateral, by Definition 15, since it has three right angles, $\angle JAB$, $\angle ABK$, and $\angle BKJ$.

In $LQKBAJ$ and $LQEFGH$, $(J,A) \cong (H,E)$ and $(A,B) \cong (E,F)$. Hence, by Theorem 35, $LQKBAJ \cong LQEFGH$. Thus, $\angle J \cong \angle H$ and $(J,K) \cong (H,G)$.

Since ∠D ≅ ∠H and ∠J ≅ ∠H, then, by Part II material (Axiom 16), ∠J ≅ ∠D.

Therefore, ⌐KCDJ is a Saccheri quadrilateral, by Theorem 31, since ∠K and ∠C are right angles and ∠D ≅ ∠J. By Definition 14, (K,J) ≅ (C,D).

Since (J,K) ≅ (H,G) and (K,J) ≅ (C,D), then, by Part II material (Axiom 13), (D,C) ≅ (H,G).

In **L⌐**ABCD and **L⌐**EFGH, ∠D ≅ ∠H and (D,C) ≅ (H,G). Thus, by Theorem 36, **L⌐**ABCD ≅ **L⌐**EFGH. ∎

Theorem 38: If **S⌐**ABCD and **S⌐**EFGH are two Saccheri quadrilaterals such that ∠D ≅ ∠H and (A,B) ≅ (E,F), then **S⌐**ABCD ≅ **S⌐**EFGH.

Proof: Suppose **S⌐**ABCD and **S⌐**EFGH are two Saccheri quadrilaterals such that ∠D ≅ ∠H and (A,B) ≅ (E,F).

By Theorem 25, ∠D ≅ ∠C and ∠H ≅ ∠G. Thus, by Part II material (Axiom 16), ∠C ≅ ∠G.

By Definition 14, ∠A, ∠B, ∠E, ∠F are right angles.

By Part II material (Theorem 88), there exist midpoints J and K for (A,B) and (D,C), respectively. Thus, A•J•B, (A,J) ≅ (J,B), D•K•C, and (D,K) ≅ (K,C).

By Theorem 24, J ≠ K; hence, line JK.

By Theorem 23, JK is perpendicular to each of (A,B) and (D,C). Hence, ∠AJK and ∠DKJ are right angles. By Part II material (Theorem 74), ∠BJK and ∠CKJ are right angles.

Therefore, ⌐AJKD and ⌐BJKC are Lambert quadrilaterals since each has three right angles. (See **Figure 11.52**.)

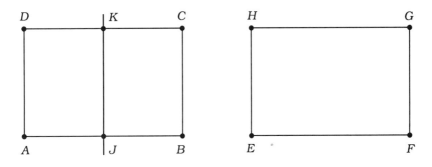

Figure 11.52

Since $A \bullet J \bullet B$ and $(A,B) \cong (E,F)$, then, by Part II material (namely, Theorem 48), there exists exactly one point M such that $E \bullet M \bullet F$ and $(A,J) \cong (E,M)$.

Since $A \bullet J \bullet B$, $E \bullet M \bullet F$, $(A,B) \cong (E,F)$, and $(A,J) \cong (E,M)$, then, by Part II material (Theorem 47), $(J,B) \cong (M,F)$.

Thus, $(A,J) \cong (J,B)$, $(A,J) \cong (E,M)$, and $(J,B) \cong (M,F)$. By Part II material (Axiom 13), $(E,M) \cong (M,F)$. Hence, M is the midpoint of (E,F).

By Part II material (Theorem 88), there exists a midpoint P for (H,G). By Theorem 24, $M \neq P$; hence, line MP.

By Theorem 23, MP is perpendicular to each of (E,F) and (H,G). Hence, $\angle EMP$ and $\angle HPM$ are right angles. By Part II material (Theorem 74), $\angle FMP$ and $\angle GPM$ are right angles. (See **Figure 11.53**.)

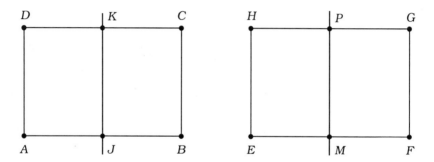

Figure 11.53

Therefore, **Q**EMPH and **Q**FMPG are Lambert quadrilaterals, since each has three right angles.

In **LQ**AJKD and **LQ**EMPH, $\angle D \cong \angle H$ and $(A,J) \cong (E,M)$. Thus, by Theorem 37, **LQ**AJKD $\cong$ **LQ**EMPH. Hence, by Definition 16, $(A,D) \cong (E,H)$ and $(D,K) \cong (H,P)$.

In **LQ**BJKC and **LQ**FMPG, $\angle C \cong \angle G$ and $(J,B) \cong (M,F)$. Thus, by Theorem 37, **LQ**BJKC $\cong$ **LQ**FMPG. Hence, by Definition 16, $(B,C) \cong (F,G)$ and $(K,C) \cong (P,G)$.

Since $D \bullet K \bullet C$, $H \bullet P \bullet G$, $(D,K) \cong (H,P)$, and $(K,C) \cong (P,G)$, then, by Part II material (Axiom 14), $(D,C) \cong (H,G)$.

Hence, in **SQ**ABCD and **SQ**EFGH, $(A,B) \cong (E,F)$, $(B,C) \cong (F,G)$, $(D,C) \cong (H,G)$, $(A,D) \cong (E,H)$, $\angle D \cong \angle H$, and $\angle C \cong \angle G$. Thus, by Definition

16. $SQABCD \cong SQEFGH$. ∎

Theorem 39: If $SQABCD$ and $SQEFGH$ are two Saccheri quadrilaterals such that $\angle D \cong \angle H$ and $(B,C) \cong (F,G)$, then $SQABCD \cong SQEFGH$.

Theorem 40: If $SQABCD$ and $SQEFGH$ are two Saccheri quadrilaterals such the $\angle D \cong \angle H$ and $(D,C) \cong (H,G)$, then $SQABCD \cong SQEFGH$.

Theorem 41: If $SQABCD$ and $SQEFGH$ are two Saccheri quadrilaterals such that $(A,B) \cong (E,F)$ and $(A,D) \cong (E,H)$, then $SQABCD \cong SQEFGH$.

Exercises 11.4

1. Develop a proof for Theorem 23.
2. Complete the proof for Theorem 26.
3. Complete the proof for Theorem 27.
4. Develop a proof for Theorem 28.
5. Develop a proof for Theorem 29.
6. Complete the proof for Theorem 30.
7. Complete the proof for Theorem 32.
8. Complete the proof for Theorem 33.
9. Develop a proof for Theorem 34.
10. Develop a proof for Theorem 35.
11. Complete the proof for Theorem 36.
12. Develop a proof for Theorem 39.
13. Develop a proof for Theorem 40.
14. Develop a proof for Theorem 41.

APPENDICES

1. Axioms and Theorems of Part II
2. Some Plane Geometric Constructions
3. Some Euclidean Theorems
4. Some Fundamental Locus Theorems
5. Some Area and Perimeter Formulas

Appendix 1

Axioms and Theorems of Part II

Axioms:

Axiom 0: There exists exactly one plane, which contains all points. (p. 77)

Axiom 1a: If A and B are two points, then there exists at least one line containing them. (p. 77)

Axiom 1b: If A and B are two points, then there exists at most one line containing them. (p. 77)

Axiom 2: If m is a line, then it contains at least two points. (p. 77)

Axiom 3: If m is a line, then there exists at least one point which does not lie on it. (p. 77)

Axiom 4: There exist at least two points. (p. 77)

Axiom 5: If A and B are two points, then
(a)   there exists at least one point C such that $A \bullet C \bullet B$;
(b)   there exists at least one point D such that $A \bullet B \bullet D$;
(c)   there exists at least one point E such that $E \bullet A \bullet B$. (p. 78)

Axiom 6: If A, B, C are points such that $A \bullet B \bullet C$, then A, B, C are distinct and collinear. (p. 79)

Axiom 7: If A, B, C are points such that $A \bullet B \bullet C$, then $C \bullet B \bullet A$. (p. 79)

Axiom 8: If A, B, C are three collinear points, then exactly one of the following is true:
(a) $A \bullet B \bullet C$;
(b) $A \bullet C \bullet B$;
(c) $C \bullet A \bullet B$. (p. 79)

Axiom 9: If A, B, C, D are points such that $A \bullet B \bullet C \bullet D$, then $A \bullet B \bullet C$, $A \bullet B \bullet D$, $A \bullet C \bullet D$, and $B \bullet C \bullet D$. (Analogous statements can be made for more than four points.) (p. 79)

Axiom 10: If A, B, C, D are four collinear points and $A \bullet B \bullet C$, then exactly one of the following is true:
(a) $A \bullet B \bullet C \bullet D$;
(b) $A \bullet B \bullet D \bullet C$;
(c) $A \bullet D \bullet B \bullet C$;
(d) $D \bullet A \bullet B \bullet C$. (p. 79)

Axiom 11: If A, B, C are three noncollinear points, m is a line which does not contain any of A, B, C, and m passes through a point of segment (A,B), then m also passes through a point of segment (B,C) or a point of segment (A,C). (p. 87)

Axiom 12: If A and B are two points on a line m and C is a point on a line n, then there exist on n exactly two points D and E, one on each side of C, such that $(A,B) \cong (C,D)$ and $(A,B) \cong (C,E)$. (p. 124)

Axiom 13: (a)   If (A,B) is a segment, then $(A,B) \cong (A,B)$.

(b)   If (A,B) and (C,D) are segments such that $(A,B) \cong (C,D)$, then $(C,D) \cong (A,B)$.

(c)   If (A,B), (C,D), (E,F) are segments such that $(A,B) \cong (C,D)$ and $(C,D) \cong (E,F)$, then $(A,B) \cong (E,F)$. (p. 124)

Axiom 14: If A, B, C, D, E, F are points such that $A \bullet B \bullet C$, $D \bullet E \bullet F$, $(A,B) \cong (D,E)$, and $(B,C) \cong (E,F)$, then $(A,C) \cong (D,F)$. (p. 125)

Axiom 15: If $\angle ABC$ is an angle and $\overrightarrow{DE}$ is a halfline on a line m, then there exist exactly two halflines $\overrightarrow{DF}$ and $\overrightarrow{DG}$, one on each side of m, such that $\angle ABC \cong \angle EDF$ and $\angle ABC \cong \angle EDG$. (p. 126)

Axiom 16: (a)   If $\angle ABC$ is an angle, then $\angle ABC \cong \angle ABC$.

(b)   If $\angle ABC$ and $\angle DEF$ are angles such $\angle ABC \cong \angle DEF$, then $\angle DEF \cong \angle ABC$.

(c)   If $\angle ABC$, $\angle DEF$, $\angle GHI$ are angles such that $\angle ABC \cong \angle DEF$ and $\angle DEF \cong \angle GHI$, then $\angle ABC \cong \angle GHI$. (p. 127)

Axiom 17: If $\triangle ABC$ and $\triangle DEF$ are triangles such that $(A,B) \cong (D,E)$, $\angle B \cong \angle E$, and $(B,C) \cong (E,F)$, then $\angle A \cong \angle D$ and $\angle C \cong \angle F$. (p. 127)

Axiom 18: If m is a line and A is a point which is not on m, then there exists exactly one line which contains A and is parallel to m. (p. 162)

Axiom 19a: If A, B, C are three noncollinear points, then there exists at least one plane containing them. (p. 182)

Axiom 19b: If A, B, C are three noncollinear points, then there exists at most one plane containing them. (p. 183)

Axiom 20: If P is a plane, then it contains at least three noncollinear points. (p. 183)

Axiom 21: If at least two points of a line m lie on a plane P, then every point lying on m lies on P. (p. 183)

Axiom 22: If two planes have a point in common, then they have at least one more point in common. (p. 183)

Axiom 23: If P is a plane, then there exists at least one point which does not lie on it. (p. 183)

Theorems:

Theorem 1: There exist at least three points. (p. 77)

Theorem 2: Two lines have at most one point in common. (p. 77)

Theorem 3: If A and B are two points on a line m and C is a point not on m, then A, B, C are noncollinear and there exist lines n and k, containing A, C and B, C, respectively, which are distinct from m and from each other. (p. 77)

Theorem 4: There exist at least two lines through any point. (p. 78)

Theorem 5: If A is a point, then there exists at least one line which does not contain it. (p. 78)

Theorem 6: If A, B, C, D are points such that $A \bullet B \bullet C$ and $B \bullet C \bullet D$, then $A \bullet B \bullet C \bullet D$. (p. 80)

Theorem 7: If A, B, C, D are points such that $A \bullet B \bullet C$ and $A \bullet C \bullet D$, then $A \bullet B \bullet C \bullet D$. (p. 80)

Theorem 8: If A, B, C, D are points such that $A \bullet B \bullet D$ and $B \bullet C \bullet D$, then $A \bullet B \bullet C \bullet D$. (p. 81)

Theorem 9: If A, B, C, D are four points such that $A \bullet B \bullet C$ and $A \bullet B \bullet D$, then exactly one of $A \bullet C \bullet D$ and $A \bullet D \bullet C$ is true. (p. 81)

Theorem 10: If A, B, C, D are four points such that $A \bullet B \bullet C$ and $A \bullet B \bullet D$, then exactly one of $B \bullet C \bullet D$ and $B \bullet D \bullet C$ is true. (p. 81)

Theorem 11: If A, B, C, D are four points such that $A \bullet B \bullet D$ and $A \bullet C \bullet D$, then exactly one of $A \bullet B \bullet C$ and $A \bullet C \bullet B$ is true. (p. 81)

Theorem 12: If m is a line and O, A are two points on m, then $\{O\}$ and L_1, L_2, the halflines of m with respect to O relative to A, partition m. (p. 83)

Theorem 13: If O is a point of a line m, then $\{O\}$ separates the two halflines which O determines on m. (p. 85)

Theorem 14: If A and B are two points, then $(A,B) = (B,A)$. (p. 86)

Theorem 15: If A and B are two points, then (A,B) is contained in line AB. (p. 87)

Theorem 16: If C is a point on a segment (A,B), then $\{C\}$, (A,C), and (C,B) partition (A,B). (p. 87)

Theorem 17: If C is a point on a segment (A,B), then $\{C\}$ separates (A,C) and (C,B). (p. 87)

Theorem 18: If A, B, C are three noncollinear points and m is a line which passes through a point of segment (A,B) and a point of segment (B,C), then m cannot pass through a point of segment (A,C). (p. 87)

Theorem 19: If m is a line on plane P and A is a point of P which is not on m, then m and P_1, P_2, the halfplanes of P with respect to m relative to A, partition **P**. (p. 90)

Theorem 20: If m is a line on plane **P**, then m separates the two halfplanes which it determines on **P**. (p. 90)

Theorem 21: If a halfline has its endpoint on a line m but does not lie on m, then all the points of the halfline lie on the same side of m. (p. 90)

Theorem 22: If m is a line and A, B, and C are three points not on m, then

(a) if A and B are on the same side of m and B and C are on the same side of m, then A and C are on the same side of m;

(b) if A and B are on the same side of m and B and C are on opposite sides of m, then A and C are on opposite sides of m;

(c) if A and B are on opposite sides of m and B and C are on opposite sides of m, then A and C are on the same side of m. (p. 91)

Theorem 23: There exist infinitely many distinct points on any segment. (p. 92)

Theorem 24: There exist infinitely many distinct points on any halfline. (p. 92)

Theorem 25: There exist infinitely many distinct points on any line. (p. 92)

Theorem 26: There exist infinitely many distinct lines through any point. (p. 92)

Theorem 27: If A is a point, then there exist infinitely many distinct lines which do not contain A. (p. 92)

Theorem 28: Every line is convex. (p. 93)

Theorem 29: Each halfline of a line m, with respect to a point O on m, is convex. (p. 93)

Theorem 30: Every segment is convex. (p. 93)

Theorem 31: Each halfplane of a plane **P**, with respect to a line m on **P**, is convex. (p. 93)

Theorem 32: The intersection of a finite number n, $n \geq 2$, of convex sets is convex. (p. 93)

Theorem 33: If D and E are points on two sides of a triangle $\triangle ABC$ and F is a point such that $D \bullet E \bullet F$, then F is in the $int(\triangle ABC)$. (p. 104)

Theorem 34: The interior of a triangle is nonempty. (p. 106)

Theorem 35: If D and E are points on two sides of a triangle $\triangle ABC$ and F is a point on line DE and in $int(\triangle ABC)$, then $D \bullet F \bullet E$. (p. 106)

Theorem 36: If a line passes through a vertex of a triangle and contains a point in the triangle's interior, then it intersects the side opposite that vertex. (p. 106)

Theorem 37: The interior of an angle is nonempty. (p. 108)

Theorem 38: If A and B are points on two sides of an angle $\angle O$ and C

is a point such that $A \bullet C \bullet B$, then C is in $int(\angle O)$. (p. 108)

Theorem 39: If A and B are points on two sides of an angle $\angle O$ and C is a point on line AB and in $int(\angle O)$, then $A \bullet C \bullet B$. (p. 109)

Theorem 40: If a point C lies in the interior of an angle $\angle AOB$, then the halfline $\overrightarrow{OC}$ is a subset of $int(\angle AOB)$. (p. 109)

Theorem 41: If point C lies in the interior of an angle $\angle AOB$, then halfline $\overrightarrow{OC}$ intersects the segment (A,B). (p. 109)

Theorem 42: If a point C lies in the interior of an angle $\angle AOB$ and a point D lies in the exterior of $\angle AOB$, then the segment (C,D) intersects $\angle AOB$. (p. 109)

Theorem 43: If $\overrightarrow{OA}$ and $\overrightarrow{OB}$ are noncollinear halflines, then a third halfline $\overrightarrow{OC}$ is between $\overrightarrow{OA}$ and $\overrightarrow{OB}$ if and only if there exist points A^*, C^*, B^* on $\overrightarrow{OA}$, $\overrightarrow{OC}$, $\overrightarrow{OB}$, respectively, such that $A^* \bullet C^* \bullet B^*$. (p. 114)

Theorem 44: If $\overrightarrow{OA}$, $\overrightarrow{OB}$, $\overrightarrow{OC}$, $\overrightarrow{OD}$ are halflines, with $\overrightarrow{OA}$ and $\overrightarrow{OD}$ being opposite halflines, and $\overrightarrow{OB}$ is between $\overrightarrow{OA}$ and $\overrightarrow{OC}$, then $\overrightarrow{OC}$ is between $\overrightarrow{OB}$ and $\overrightarrow{OD}$. (p. 114)

Theorem 45: If $\overrightarrow{OA}$, $\overrightarrow{OB}$, $\overrightarrow{OC}$, $\overrightarrow{OD}$ are halflines, with $\overrightarrow{OA}$, $\overrightarrow{OB}$ being noncollinear, such that $\overrightarrow{OC}$ is between $\overrightarrow{OA}$, $\overrightarrow{OB}$ and $\overrightarrow{OD}$ is between $\overrightarrow{OB}$, $\overrightarrow{OC}$, then $\overrightarrow{OD}$ is between $\overrightarrow{OA}$ and $\overrightarrow{OB}$. (p. 114)

Theorem 46: If $\overrightarrow{OA}$, $\overrightarrow{OB}$, $\overrightarrow{OC}$, $\overrightarrow{OD}$ are halflines, with $\overrightarrow{OA}$, $\overrightarrow{OB}$ noncollinear, $\overrightarrow{OC}$, $\overrightarrow{OD}$ noncollinear, D on the C-side of $\overrightarrow{OA}$, such that $\overrightarrow{OC}$ is between $\overrightarrow{OA}$, $\overrightarrow{OB}$ and $\overrightarrow{OB}$ between $\overrightarrow{OC}$, $\overrightarrow{OD}$, then $\overrightarrow{OB}$ is between $\overrightarrow{OA}$ and $\overrightarrow{OD}$. (p. 114)

Theorem 47: If A, B, C, D, E, F are points such that $A \bullet B \bullet C$, $D \bullet E \bullet F$, $(A,B) \cong (D,E)$, and $(A,C) \cong (D,F)$, then $(B,C) \cong (E,F)$. (p. 125)

Theorem 48: If A, B, C, D, E are points such that $A \bullet B \bullet C$ and $(A,C) \cong (D,E)$, then there exists a exactly one point F such that $(A,B) \cong (D,F)$

and $D \bullet F \bullet E$. (p. 126)

Theorem 49: (a)   If $\triangle ABC$ is a triangle, then $\triangle ABC \cong \triangle ABC$.
(b)   If $\triangle ABC$ and $\triangle DEF$ are triangles such that $\triangle ABC \cong \triangle DEF$, then $\triangle DEF \cong \triangle ABC$.
(c)   If $\triangle ABC$, $\triangle DEF$, $\triangle GHI$ are triangles such that $\triangle ABC \cong \triangle DEF$ and $\triangle DEF \cong \triangle GHI$, then $\triangle ABC \cong \triangle GHI$. (p. 127)

Theorem 50: If $\triangle ABC$ and $\triangle DEF$ are triangles such that $(A,B) \cong (D,E)$, $\angle B \cong \angle E$, and $(B,C) \cong (E,F)$, then $\triangle ABC \cong \triangle DEF$. (p. 128)

Theorem 51: If $\triangle ABC$ is a triangle such that $(A,B) \cong (A,C)$, then $\angle B \cong \angle C$. (p. 129)

Theorem 52: If $\angle ABC \cong \angle DEF$ and halfline $\overrightarrow{BG}$ is between halflines $\overrightarrow{BA}$ and $\overrightarrow{BC}$, then there exists a unique halfline $\overrightarrow{EH}$ such that $\overrightarrow{EH}$ is between halflines $\overrightarrow{ED}$ and $\overrightarrow{EF}$, $\angle ABG \cong \angle DEH$, and $\angle GBC \cong \angle HEF$. (p. 129)

Theorem 53: If $\angle ABC$, $\angle DEF$, $\angle ABG$, $\angle DEH$, $\angle GBC$, $\angle HEF$ are angles such that halfline $\overrightarrow{BG}$ is between halflines $\overrightarrow{BA}$ and $\overrightarrow{BC}$, halfline $\overrightarrow{EH}$ is between halflines $\overrightarrow{ED}$ and $\overrightarrow{EF}$, $\angle ABG \cong \angle DEH$, and $\angle GBC \cong \angle HEF$, then $\angle ABC \cong \angle DEF$. (p. 131)

Theorem 54: If $\angle ABC$, $\angle DEF$, $\angle ABG$, $\angle DEH$, $\angle GBC$, $\angle HEF$ are angles such that halfline $\overrightarrow{BG}$ is between halflines $\overrightarrow{BA}$ and $\overrightarrow{BC}$, halfline $\overrightarrow{EH}$ is between halflines $\overrightarrow{ED}$ and $\overrightarrow{EF}$, $\angle ABC \cong \angle DEF$, and $\angle ABG \cong \angle DEH$, then $\angle GBC \cong \angle HEF$. (p. 131)

Theorem 55: If $\triangle ABC$ and $\triangle DEF$ are triangles such that $\angle A \cong \angle D$, $(A,C) \cong (D,F)$, and $\angle C \cong \angle F$, then $\triangle ABC \cong \triangle DEF$. (p. 131)

Theorem 56: If $\triangle ABC$ and $\triangle DEF$ are two triangles such that $(A,B) \cong (D,E)$, $(B,C) \cong (E,F)$, and $(A,C) \cong (D,F)$, then $\triangle ABC \cong \triangle DEF$. (p. 132)

Theorem 57: If (A,B), (C,D), (E,F) are segments such that $(A,B) < (C,D)$ and $(A,B) \cong (E,F)$, then $(E,F) < (C,D)$. (p. 135)

Theorem 58: If (A,B), (C,D), (E,F) are segments such that $(A,B) < (C,D)$ and $(C,D) \cong (E,F)$, then $(A,B) < (E,F)$. (p. 135)

Theorem 59: If (A,B), (C,D), (E,F) are segments such that $(A,B) < (C,D)$ and $(C,D) < (E,F)$, then $(A,B) < (E,F)$. (p. 135)

Theorem 60: If (A,B) and (C,D) are two segments, then exactly one of the following is true:

(a) $(A,B) < (C,D)$;

(b) $(A,B) \cong (C,D)$;

(c) $(C,D) < (A,B)$. (p. 135)

Theorem 61: If ∠ABC, ∠DEF, ∠GHI are angles such that ∠ABC < ∠DEF and ∠ABC ≅ ∠GHI, then ∠GHI < ∠DEF. (p. 138)

Theorem 62: If ∠ABC, ∠DEF, ∠GHI are angles such that ∠ABC < ∠DEF and ∠DEF ≅ ∠GHI, then ∠ABC < ∠GHI. (p. 138)

Theorem 63: If ∠ABC, ∠DEF, ∠GHI are angles such that ∠ABC < ∠DEF and ∠DEF < ∠GHI, then ∠ABC < ∠GHI. (p. 138)

Theorem 64: If ∠ABC and ∠DEF are two angles, then exactly one of the following is true:

(a) ∠ABC < ∠DEF;

(b) ∠DEF < ∠ABC;

(c) ∠ABC ≅ ∠DEF. (p. 138)

Theorem 65: If two angles are congruent, then the angles with which they form linear pairs are congruent. (p. 139)

Theorem 66: If two angles are congruent, then their supplementary angles are congruent. (p. 141)

Theorem 67: If two angles are adjacent and supplementary, then they are a linear pair of angles. (p. 141)

Theorem 68: Vertical angles are congruent. (p. 142)

Theorem 69: An angle congruent to a right angle is a right angle. (p. 142)

Theorem 70: If two angles are right angles, then they are congruent. (p. 142)

Theorem 71: If one of two angles is a right angle and the other is an obtuse angle, then the right angle is less than the obtuse angle. (p. 144)

Theorem 72: If one of two angles is an acute angle and the other is an obtuse angle, then the acute angle is less than the obtuse angle. (p. 144)

Theorem 73: For a pair of supplementary angles, one is an acute angle if and only if the other is an obtuse angle. (p. 145)

Theorem 74: If two intersecting lines contain one right angle, then they contain four right angles. (p. 146)

Theorem 75: If two angles are supplementary and not congruent, then one of them is an acute angle. (p. 146)

Theorem 76: If two lines intersected by a transversal have a pair of alternate interior angles which are congruent, then the two lines are parallel. (p. 160)

Theorem 77: If two lines intersected by a transversal have two interior angles which are supplementary and have their interiors on the same side of the transversal, then the two lines are parallel. (p. 161)

Theorem 78: If two lines intersected by a transversal have a pair of corresponding angles which are congruent, then the two lines are parallel. (p. 161)

Theorem 79: If two parallel lines are intersected by a transversal, then each pair of alternate interior angles are congruent. (p. 162)

Theorem 80: If two parallel lines are intersected by a transversal, then each pair of interior angles with their interiors on the same side of the transversal are supplementary. (p. 162)

Theorem 81: If two parallel lines are intersected by a transversal, then each pair of corresponding angles are congruent. (p. 162)

Theorem 82: If a line intersects one of two parallel lines, then it intersects the other. (p. 162)

Theorem 83: If m, n, t are three lines such that each of m and n is parallel to t, then m and n are parallel. (p. 162)

Theorem 84: An interior angle of a triangle is less than each of its remote exterior angles. (p. 163)

Theorem 85: If two sides of a triangle are not congruent, then their opposite angles are not congruent and the lesser angle is opposite the lesser side. (p. 165)

Theorem 86: If two angles of a triangle are not congruent, then their opposite sides are not congruent and the lesser side is opposite the lesser angle. (p. 166)

Theorem 87: If $\triangle ABC$ and $\triangle DEF$ are triangles such that $(A,B) \cong (D,E)$, $\angle A \cong \angle D$, and $\angle C \cong \angle F$, then $\triangle ABC \cong \triangle DEF$. (p. 166)

Theorem 88: Every segment has exactly one midpoint. (p. 167)

Theorem 89: Every angle has exactly one bisector. (p. 169)

Theorem 90: If m is a line and A is a point not on m, then there exists exactly one line which contains A and is perpendicular to m. (p. 170)

Theorem 91: There exists at least one right angle. (p. 172)

Theorem 92: If m is a line and A is a point on m, then there exists exactly one line which contains A and is perpendicular to m. (p. 172)

Theorem 93: If a line is perpendicular to one of two parallel lines, it is perpendicular to the other. (p. 172)

Theorem 94: If two lines are perpendicular, respectively, to a pair of parallel lines, then the lines are parallel. (p. 172)

Theorem 95: If m is a line, A is a point not on m, B is the point on m such that (A,B) is perpendicular to m, and C is a second point on m, then $(A,B) < (A,C)$. (p. 173)

Theorem 96: At least two angles of any triangle are acute. (p. 173)

Theorem 97: Every segment has exactly one perpendicular bisector. (p. 174)

Theorem 98: A point is equidistant from the endpoints of a segment if and only if it lies on the perpendicular bisector of the segment. (p. 174)

Theorem 99: If m is a line and A is a point not on m, then there exists exactly one plane containing m and A. (p. 183)

Theorem 100: If m and n are two lines which have a point in common, then there exists exactly one plane containing them. (p. 183)

Theorem 101: Two lines lie together on at most one plane. (p. 183)

Theorem 102: Not all lines lie on the same plane. (p. 183)

Theorem 103: If A is a point on a plane P, then there exists at least one line which does not contain A and does not lie on P. (p. 183)

Theorem 104: There exist at least two planes through any point. (p. 183)

Theorem 105: If A is a point, then there exists at least one plane which does not contain it. (p. 183)

Theorem 106: If two planes have a point in common, then all their common points lie on exactly one line. (p. 183)

Theorem 107: If P is a plane and m is a line which does not lie on P, then P and m have at most one point in common. (p. 184)

Theorem 108: There exist at least six space lines. (p. 184)

Theorem 109: There exist at least four planes. (p. 184)

Theorem 110: If P is a plane and A is a point which is not on P, then P and S_1, S_2, the halfspaces with respect to P relative to A, partition space. (p. 185)

Theorem 111: If P is a plane, then it separates the two halfspaces which it determines. (p. 185)

Theorem 112: Every halfspace, with respect to a plane P, is convex. (p. 185)

Theorem 113: There exist infinitely many distinct lines on any plane. (p. 186)

Theorem 114: If A is a point on a plane P, then there exist infinitely many distinct space lines which do not contain A and do not lie on P. (p. 186)

Theorem 115: There exist infinitely many distinct planes through any point. (p. 186)

Appendix 2

Some Plane Geometric Constructions

Construction 1: To construct a line segment congruent to a given line segment.

Given: Line segment $\overline{AB}$.
To be constructed: A line segment congruent to $\overline{AB}$.
Construction: On a working line w, with any point C as a center and a radius congruent to $\overline{AB}$, construct an arc intersecting w at D. $\overline{CD}$ is the required line segment.

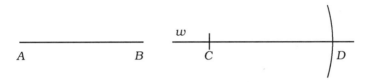

Construction 2: To construct an angle congruent to a given angle.

Given: Angle $\angle A$.
To be constructed: An angle congruent to $\angle A$.
Construction:
(1) With A as center and using a convenient radius, construct an arc a_1 intersecting the sides of $\angle A$ at B and C.
(2) With A', a point on working line w, as center and using the same radius, construct arc a_2 intersecting w at B'.
(3) With B' as center and using a radius congruent to $\overline{BC}$, construct arc a_3 intersecting arc a_2 at C'.
(4) Draw $\overrightarrow{A'C'}$. $\angle A'$ is the required angle.

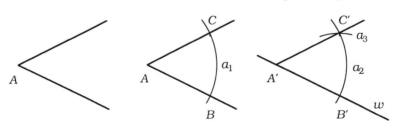

Construction 3: To bisect a given angle.

> Given: Angle ∠A.
> To be constructed: The bisector of ∠A.
> Construction: (1) With A as center and using a convenient radius, construct an arc intersecting the sides of ∠A at B and C.
> (2) With B and C as centers and using equal radii, construct arcs intersecting at D.
> (3) Draw ray $\overrightarrow{AD}$. $\overrightarrow{AD}$ is the required bisector.

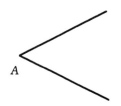
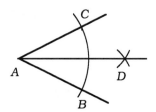

Construction 4: To construct a line perpendicular to a given line through a given point on the line.

> Given: Line w and point P on w.
> To be constructed: A perpendicular to w at P.
> Construction: Bisect the straight angle at P. Line DP is the required perpendicular.

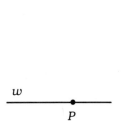
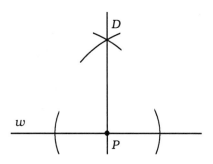

Construction 5: To bisect a given line segment. (To construct the perpendicular bisector of a given line segment.)

> Given: Line segment $\overline{AB}$.
> To be constructed: Perpendicular bisector of $\overline{AB}$.
> Construction: (1) With A as center and using a radius of more than

half of $\overline{AB}$, construct arc a_1.
(2) With B as center and using the same radius, construct arc a_2 intersecting arc a_1 at C and D.
(3) Draw line CD. Line CD is the required perpendicular bisector of $\overline{AB}$.

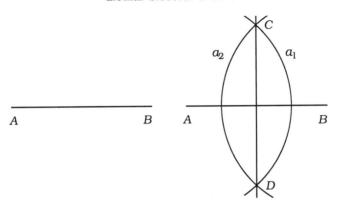

Construction 6: To construct a line perpendicular to a given line through a given external point.

Given: Line w and point P not on w.
To be constructed: A perpendicular to w through P.
Construction: (1) With P as center and using a sufficiently large radius, construct an arc intersecting w at B and C.
(2) With B and C as centers and using equal radii which are more than half of $\overline{BC}$, construct arcs intersecting at A.
(3) Draw line PA. PA is the required perpendicular.

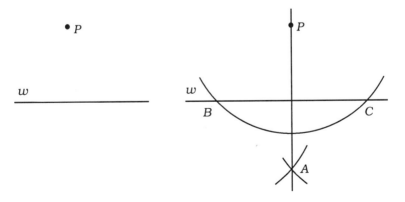

Construction 7: To construct a triangle given its three sides.

Given: Sides a, b, and c of $\triangle ABC$.
To be constructed: $\triangle ABC$.
Construction: (1) On a working line w, construct $\overline{AC} \cong b$.
(2) With A as center and c as radius, construct arc a_1.
(3) With C as center and a as radius, construct arc a_2 intersecting arc a_1 at B.
(4) Draw $\overline{BC}$ and $\overline{AB}$. $\triangle ABC$ is the required triangle.

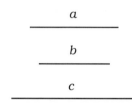
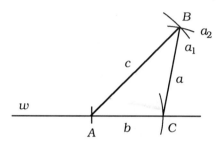

Construction 8: To construct an angle of 60°.

Given: Line w.
To be constructed: An angle of 60°.
Construction: Using a convenient length as a side, construct an equilateral triangle. Any angle of the equilateral triangle is the required angle.

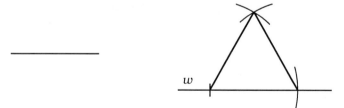

Construction 9: To construct a triangle given two sides and the included angle.

Given: $\angle A$, b, and c.
To be constructed: $\triangle ABC$.
Construction: (1) On a working line w, construct $\overline{AC} \cong b$.
(2) At A, construct $\angle A$ with $\overline{AC}$ on one side.

Appendix 2

Some Plane Geometric Constructions

Construction 1: To construct a line segment congruent to a given line segment.

Given: Line segment $\overline{AB}$.
To be constructed: A line segment congruent to $\overline{AB}$.
Construction: On a working line w, with any point C as a center and a radius congruent to $\overline{AB}$, construct an arc intersecting w at D. $\overline{CD}$ is the required line segment.

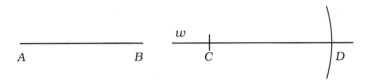

Construction 2: To construct an angle congruent to a given angle.

Given: Angle $\angle A$.
To be constructed: An angle congruent to $\angle A$.
Construction: (1) With A as center and using a convenient radius, construct an arc a_1 intersecting the sides of $\angle A$ at B and C.
(2) With A', a point on working line w, as center and using the same radius, construct arc a_2 intersecting w at B'.
(3) With B' as center and using a radius congruent to $\overline{BC}$, construct arc a_3 intersecting arc a_2 at C'.
(4) Draw $\overrightarrow{A'C'}$. $\angle A'$ is the required angle.

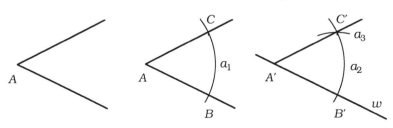

Construction 3: To bisect a given angle.

    Given: Angle ∠A.
    To be constructed: The bisector of ∠A.
    Construction: (1) With A as center and using a convenient radius, construct an arc intersecting the sides of ∠A at B and C.
                       (2) With B and C as centers and using equal radii, construct arcs intersecting at D.
                       (3) Draw ray $\vec{AD}$. $\vec{AD}$ is the required bisector.

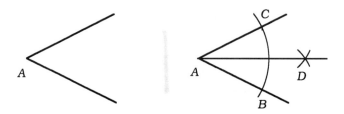

Construction 4: To construct a line perpendicular to a given line through a given point on the line.

    Given: Line w and point P on w.
    To be constructed: A perpendicular to w at P.
    Construction: Bisect the straight angle at P. Line DP is the required perpendicular.

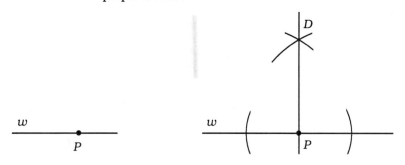

Construction 5: To bisect a given line segment. (To construct the perpendicular bisector of a given line segment.)

    Given: Line segment $\overline{AB}$.
    To be constructed: Perpendicular bisector of $\overline{AB}$.
    Construction: (1) With A as center and using a radius of more than

(3) On the other side of ∠A, construct $\overline{AB} \cong c$.
(4) Draw $\overline{BC}$. △ABC is the required triangle.

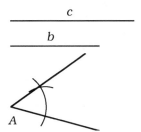
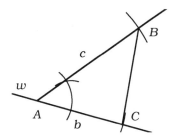

Construction 10: To construct a triangle given two angles and the included side.

Given: ∠A, ∠C and b.
To be constructed: △ABC.

Construction:
(1) On a working line w, construct $\overline{AC} \cong b$.
(2) At A construct ∠A with $\overline{AC}$ on one side, and at C construct ∠C with $\overline{AC}$ on one side.
(3) Extend the new sides of the angles until they meet, at B. △ABC is the required triangle.

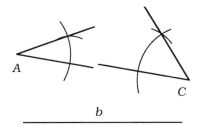
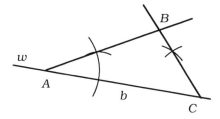

Construction 11: To construct a triangle given two angles and a side not included.

Given: ∠A, ∠B and b.
To be constructed: △ABC.

Construction:
(1) On a working line w, construct $\overline{AC} \cong b$.
(2) At C construct an angle whose measure equals m∠A + m∠B such that the extension of $\overline{AC}$ will be one side of the angle. The remainder of the straight angle at C will be ∠C.

(3) At A construct $\angle A$ with $\overline{AC}$ as one side. The intersection of the new sides of the angles is B. $\triangle ABC$ is the required triangle.

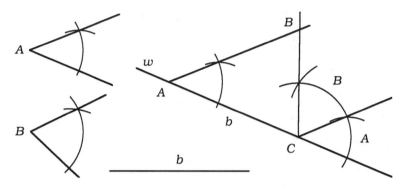

Construction 12: To construct a right triangle given its hypotenuse and a leg.

Given: Hypotenuse c and leg b of right triangle $\triangle ABC$.
To be constructed: Right triangle $\triangle ABC$.
Construction: (1) On a working line w, construct $\overline{AC} \cong b$.
(2) Construct a perpendicular to w at C.
(3) With A as center and c as radius, construct an arc intersecting the perpendicular at B. $\triangle ABC$ is the required triangle.

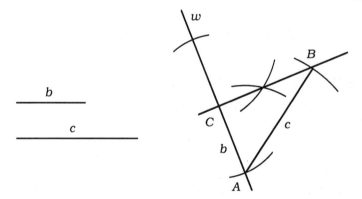

Construction 13: To construct a line parallel to a given line through a given external point.

Given: Line w and a point P not on w.

328   APPENDIX 2

To be constructed: A line through P parallel to w.
Construction: (1) Draw a line m through P intersecting w at Q.
(2) Select a third point R on line m such that P is between Q and R. Select a second point S on line w.
(3) Construct an angle which is congruent of ∠PQS such that it lies on the same side of m as does ∠PQS, has P as its vertex, and has $\overrightarrow{PR}$ as one side. The line which contains the other side of this angle is the required line. (If two corresponding angles are congruent, the lines cut by the transversal are parallel.)

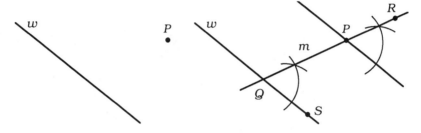

Construction 14: To construct a tangent to a given circle through a given point on the circle.

Given: Circle having center O and a point P on the circle.
To be constructed: A tangent to the circle at P.
Construction: Draw the ray $\overrightarrow{OP}$ and construct a line m perpendicular to $\overrightarrow{OP}$ at P. Line m is the required tangent. (A line perpendicular to a radius at its outer end is a tangent to the circle.)

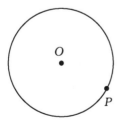 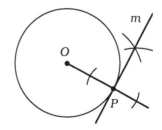

SOME PLANE GEOMETRIC CONSTRUCTIONS

Construction 15: To construct a tangent to a given circle through a given point outside the circle.

Given: Circle having center O and point P outside the circle.
To be constructed: A tangent to the circle from P.
Construction:
(1) Draw line OP.
(2) Construct a circle having center Q and the segment $\overline{OP}$ as a diameter.
(3) Label as A and B the points where the two circles intersected.
(4) Draw lines PA and PB, the required tangent lines. (∠OAP and ∠OBP are right angles, since angles inscribed in semicircles are right angles.)

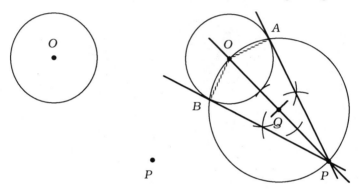

Construction 16: To circumscribe a circle about a triangle.

Given: △ABC.
To be constructed: The circumcircle of △ABC.
Construction: Construct the perpendicular bisectors of two sides of the triangle. Their intersection is the center of the required circle and the distance to any vertex is the radius. (Any point on the perpendicular bisector of a segment is equidistant from the ends of the segment.)

Construction 17: To locate the center of a given circle.

Given: A circle.
To be constructed: The center of the given circle.
Construction:
(1) Select any three points A, B and C on the circle.
(2) Construct the perpendicular bisectors of segments $\overline{AB}$ and $\overline{BC}$. The intersection of these perpendicular bisectors of the center of the circle.

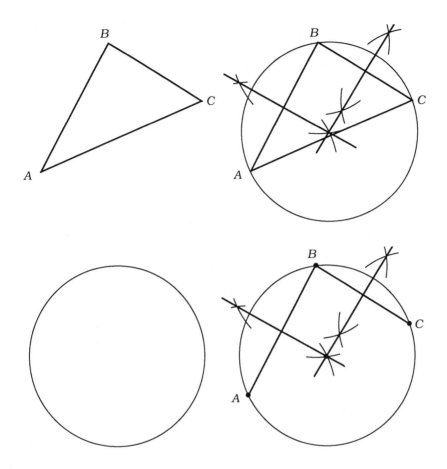

Construction 18: To inscribe a circle in a given triangle.

Given: △ABC.

To be constructed: The incircle of △ABC.

Construction: (1) Construct the bisectors of two of the angles of △ABC. Label their point of intersection as O. This point is the center of the required circle.

(2) Construct a line containing O and perpendicular to any side of △ABC. Label the point of intersection of the side and the perpendicular line as D.

Segment $\overline{OD}$ is the radius of the required circle.

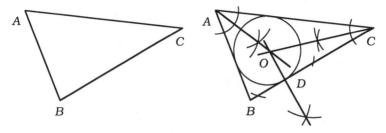

Construction 19: To divide a line segment into any number of congruent parts.

Given: Segment $\overline{AB}$.

To be constructed: Divide $\overline{AB}$ into any number of congruent parts.

Construction: (1) On a ray $\overrightarrow{AP}$, cut off the required number of congruent segments.
(2) Connect B to the endpoint, labeled C, of the last segment, and construct parallels to $\overline{BC}$. The points of intersection of these parallels and $\overline{AB}$ divide $\overline{AB}$ into the required number of segments. (If three or more parallel lines cut off congruent segments on one transversal, they cut off congruent segments on any other transversal.)

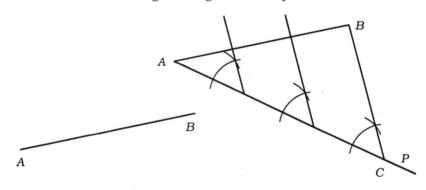

Construction 20: To divide a given line segment into parts having a given ratio.

Given: Segment $\overline{AB}$, m, and n.

To be constructed: Divide $\overline{AB}$ into two segments whose ratio is $m:n$.

Construction: (1) On a ray $\overrightarrow{AG}$, construct $\overline{AC} \cong m$ and $\overline{CD} \cong n$.

(2) Draw $\overline{BD}$ and construct $\overrightarrow{CE}$ parallel to $\overline{BD}$. $\overline{AE}$ and $\overline{EB}$ are the required segments. (A line parallel to one side of a triangle divides the other two sides into proportional segments.)

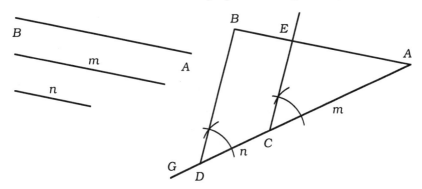

Construction 21: To construct the fourth proportional to three given line segments.

Given: Segments a, b, and c.
To be constructed: The fourth proportional to a, b, c.

Construction: (1) On a working ray $\overrightarrow{AP}$, construct $\overline{AD} \cong a$ and $\overline{DB} \cong b$.

(2) On another ray $\overrightarrow{AQ}$, construct $\overline{AE} \cong c$.

(3) Draw $\overline{DE}$ and construct $\overline{BC}$ parallel to $\overline{DE}$. $\overline{EC}$ is the required segment.

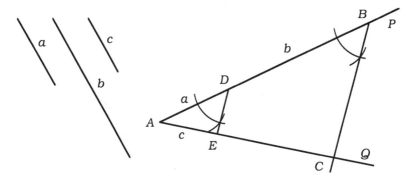

Construction 22: To construct the mean proportional to two given line segments.

Given: Segments a and b.
To be constructed: The mean proportional to a and b.
Construction: (1) On a working line w, construct $\overline{AB} \cong a$ and $\overline{BC} \cong b$.
(2) Construct a semicircular arc, using $\overline{AC}$ as diameter.
(3) At B construct a perpendicular to AC, meeting the semicircle at D. $\overline{DB}$ is the required segment. (In a right triangle, the altitude to the hypotenuse is the mean proportional between the segments of the hypotenuse.)

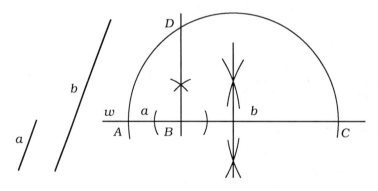

Construction 23: To inscribe a square in a given circle.

Given: Circle.
To be constructed: A square inscribed in the given circle.
Construction: Construct a diameter and construct another diameter perpendicular to it. Join consecutive endpoints of the diameters to form the required square.

Construction 24: To inscribe a regular octagon in a given circle.

Given: Circle.
To be constructed: A regular octagon inscribed in the given circle.
Construction: (1) As in Construction 23, construct perpendicular diameters.
(2) Bisect the angles formed by these diameters, producing eight congruent arcs. The chords of these arcs are the sides of the required octagon.

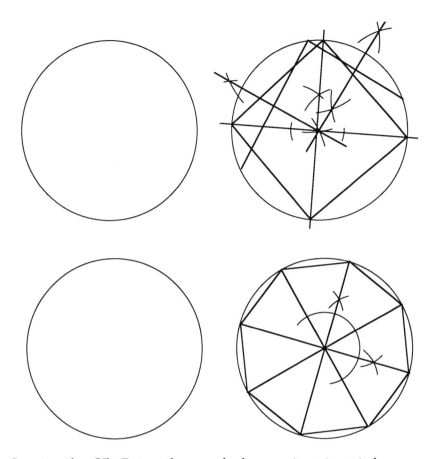

Construction 25: To inscribe a regular hexagon in a given circle.

    Given: Circle.
    To be constructed: A regular hexagon inscribed in the given circle.
    Construction: (1) Construct diameter $\overline{AD}$.
                     (2) Using A and D as centers, construct arcs having the same radius as the given circle.
                     (3) Construct the required regular hexagon by joining consecutive points in which these arcs intersect the circle.

Construction 26: To inscribe an equilateral triangle in a given circle.

    Given: Circle.
    To be constructed: An equilateral triangle inscribed in the given circle.
    Construction: Select three nonconsecutive points from the six obtained

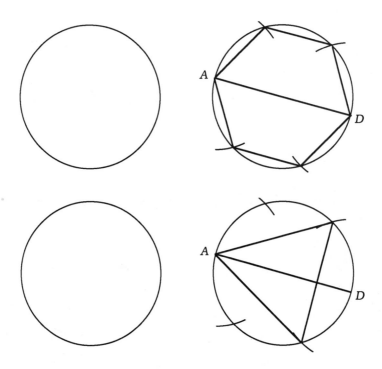

using Construction 25. Join these three points to obtain an equilateral triangle.

Appendix 3

Some Euclidean Theorems

Perpendicular Bisectors

1. Two points each equidistant from the endpoints of a segment determine the perpendicular bisector of the segment.
2. The perpendicular bisector of a segment is the set of points equidistant from the ends of the segment.
3. A diameter bisects a chord of a circle if and only if it is is perpendicular to it.
4. If two circles intersect, the line of centers is the perpendicular bisector of the common chord.
5. A radius to the midpoint of an arc is the perpendicular bisector of the chord which subtends the arc.
6. Each diagonal of a rhombus, or of a square, is the perpendicular bisector of the other diagonal.

Triangles

7. Two right triangles are congruent if any two parts, at least one of which is a side, in one are congruent to corresponding parts of the other.
8. Two right triangles are similar if an acute angle of one is congruent to an acute angle of the other.
9. In a right triangle, the altitude to the hypotenuse divides the triangle into two similar right triangles, each similar to the original triangle.
10. In a right triangle, the altitude to the hypotenuse is the mean proportional between the segments of the hypotenuse.
11. In a right triangle, if the altitude to the hypotenuse is drawn, each leg is the mean proportional between the whole hypotenuse and the adjacent segment of the hypotenuse.
12. The sum of the squares of the two legs of a right triangle equals the square of the hypotenuse.
13. The midpoint of the hypotenuse of a right triangle is equidistant from the three vertices, and is, therefore, the center of the circumscribed circle.
14. In a 30°– 60° right triangle, the side opposite the 30° angle is half the length of the hypotenuse and the side opposite the 60° angle is $\sqrt{3}/2$ times the length of the hypotenuse.

15. An angle inscribed in a semicircle forms a right triangle with the diameter which determines the semicircle.
16. The measure of an exterior angle of a triangle is greater than the measure of each remote interior angle.
17. The sum of the measures of any two of a triangle's angles is less than 180°.
18. If the lengths of two sides of a triangle are unequal, then the measures of the angles opposite those sides are unequal in the same order.
19. If the measures of two angles of a triangle are unequal, then the lengths of the sides opposite those angles are unequal in the same order.
20. Two angles of a triangle are congruent if and only if the sides opposite those angles are congruent.
21. The sum of the lengths of any two sides of a triangle is greater than the length of the third side.
22. If two sides of one triangle are congruent to two sides of another triangle but the measures of the included angles are unequal, then the lengths of the third sides are unequal in the same order.
23. If two sides of one triangle are congruent to two sides of another triangle, but the lengths of the third sides are unequal, then the measures of the angles included between the pairs of congruent sides are unequal in the same order.
24. The medians of a triangle intersect in a point, the centroid, that is two-thirds of the distance from each vertex to the midpoint of the opposite side.
25. The hypotenuse of a right triangle is a diameter of the circumscribed circle.

Parallels

26. Two lines intersected by a transversal are parallel if and only if alternate interior angles are congruent.
27. Two lines intersected by a transversal are parallel if and only if corresponding angles are congruent.
28. Two lines intersected by a transversal are parallel if and only if the interior angles on the same side of the transversal are supplementary.
29. In a parallelogram the opposite sides are congruent, the opposite angles are congruent, and the diagonals bisect each other.
30. If a series of parallel lines cut off congruent segments on one transversal, they cut off congruent segments on any transversal.
31. The segments cut off on two transversals by a series of parallel lines are proportional.

32. Two parallel lines cut off congruent arcs on a circle.
33. A line cuts two sides of a triangle proportionally if and only if it is parallel to the third side.
34. A line connecting the midpoints of two sides of a triangle is parallel to the third side and half the length of the third side.
35. If two sides of a quadrilateral are parallel and congruent, the quadrilateral is a parallelogram.
36. If both pairs of opposite sides of a quadrilateral are congruent, the quadrilateral is a parallelogram.
37. If the diagonals of a quadrilateral bisect each other, the quadrilateral is a parallelogram.
38. A parallelogram inscribed in a circle is a rectangle.
39. A trapezoid inscribed in a circle is isosceles.
40. The median of a trapezoid is parallel to the bases and its length is one half the sum of the lengths of the bases.

Angle Bisectors

41. The bisector of an angle is the set of points equidistant from the sides of the angle.
42. The bisector of an angle of a triangle divides the opposite side into segments which are proportional to the adjacent sides.
43. If a point on a side of a triangle divides the side into segments which are proportional to the adjacent sides, then it lies on the bisector of the opposite angle.
44. If an external bisector of an angle of a triangle intersects the opposite sideline, then the point of intersection divides the opposite side into segments which are proportional to the adjacent sides.
45. If a point on a sideline of a triangle divides the related side into segments which are proportional to the adjacent sides but does not lie on the side, then it lies on an external bisector of the opposite angle.
46. The bisector of the vertex angle of an isosceles triangle is a median and an altitude.
47. The line drawn from the intersection of two tangents to the center of a circle bisects the angle formed by the tangents.
48. A circle can be inscribed in any triangle.
49. The diagonals of a rhombus, or of a square, bisect the angles.

Circles

50. Chords in a circle, or in congruent circles, are congruent if and only if their arcs are congruent.

51. Chords are equally distant from the center of a circle, or the centers of congruent circles, if and only if they are congruent.
52. An inscribed angle, formed by two chords drawn from a point on a circle, is measured by one half the intercepted arc.
53. An angle formed by two chords intersecting within a circle is measured by one half the sum of the intercepted arcs.
54. If two chords intersect within a circle, the product of the segments of one equals the product of the segments of the other.
55. A circle may be circumscribed about any triangle or about any regular polygon.

Tangents and Secants

56. A tangent to a circle at a given point is perpendicular to the radius drawn to that point.
57. An angle formed by a tangent and a chord is measured by one half the intercepted arc.
58. Two segments tangent to a circle from an external point are congruent.
59. The angle formed by two tangents, two secants, or a tangent and a secant, is measured by one half the difference of the intercepted arcs of the circle.
60. If a tangent and a secant are drawn to a circle from a point, the tangent is the mean proportional between the whole secant and its external segment.
61. If two secants are drawn to a circle from an exterior point, the product of the lengths of one secant and its external segment is equal to the product of the lengths of the other secant and its external segment.
62. If a tangent and a secant are drawn to a circle from an exterior point of the circle, the square of the length of the tangent is equal to the product of the lengths of the secant and its external segment.

Angles and Sum of Angles

63. The sum of the three angles of a triangle is $180°$.
64. The angle formed by two lines tangent to a circle is the supplement of the central angle formed by the radii drawn to the points of tangency.
65. A quadrilateral can be inscribed in a circle if and only if its opposite angles are supplementary.
66. The sum of the angles of a polygon of n sides equals $(n-2)180°$.

67. The sum of the exterior angles of any polygon equals 360°.
68. An exterior angle of a triangle equals the sum of the two opposite interior angles.
69. An interior angle of a regular polygon having n sides measures $(n-2)180°/n$.
70. A central angle of a regular polygon of n sides measures $360°/n$.
71. An exterior angle of a regular polygon of n sides measures $360°/n$.
72. Verticals angles formed by two intersecting lines are congruent.

Similar Figures

73. Two triangles are similar if and only if their corresponding angles are congruent.
74. Two triangles are similar if and only if an angle of one is congruent to an angle of the other and the including sides are proportional.
75. Two triangles are similar if and only if their corresponding sides are proportional.
76. Two right triangles are similar if and only if an acute angle of one is congruent to an acute angle of the other.
77. Two right triangles are similar if and only if the hypotenuse and a side of one are proportional to the hypotenuse and a side of the other.
78. In similar triangles, the lengths of the altitudes from corresponding vertices are in the same ratio as the lengths of corresponding sides.
79. In similar triangles, the lengths of bisectors of corresponding angles are in the same ratio as the lengths of corresponding sides.
80. In similar triangles, the lengths of medians from corresponding vertices are in the same ratio as the lengths of corresponding sides.
81. Polygons are similar if they are composed of the same number of triangles which are similar and similarly placed.
82. Regular polygons of the same number of sides are similar.
83. Perimeters of similar polygons are proportional to any two corresponding sides of the polygons.
84. Circumferences of any two circles are proportional to their radii, or to their diameters.
85. If the non-parallel sides of a trapezoid are extended to meet, they form with the bases two similar triangles.

Areas

86. Areas of two circles are proportional to the squares of their radii.

87. Areas of parallelograms, or of triangles, of equal bases and altitudes are equal.
88. Areas of two similar polygons are to each other as the squares of any two corresponding lines.
89. The area of a regular polygon is one half the product of its perimeter and its apothem.
90. The area of a rhombus equals one half the product of its diagonals.
91. The area of a regular hexagon equals the sum of the areas of the six equilateral triangles formed by drawing the radii of the polygon.
92. The area of a right triangle is equal to one half the product of the two legs.
93. The area of an equilateral triangle is equal to one-fourth the square of a side multiplied by $\sqrt{3}$.

Special Figures

94. The base angles of an isosceles trapezoid are congruent.
95. The diagonals of an isosceles trapezoid are congruent.
96. All the angles of a rectangle are right angles.
97. The diagonals of a rectangle are congruent and bisect each other.
98. Each diagonal of a rectangle is a diameter of the circumscribed circle.
99. The diagonals of a rhombus, or a square, are perpendicular bisectors of each other.
100. The diagonals of a rhombus, or a square, bisect its angles.
101. The diagonals of a convex quadrilateral intersect.

Appendix 4

Some Fundamental Locus Theorems

Definition: Locus, in Latin, means location. The plural is loci.

Definition: The **locus of a point** is the set of all points, and only those points, which satisfy given conditions.

Theorem L.1: The locus of a point equidistant from two given points is the perpendicular bisector of the segment joining the two points. (See **Figure 1**.)

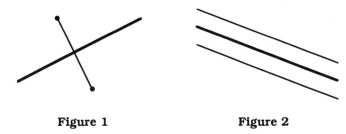

Figure 1   Figure 2

Theorem L.2: The locus of a point equidistant from two given parallel lines is a line parallel to the two lines and midway between them. (See **Figure 2**.)

Theorem L.3: The locus of a point equidistant from the sides of a given angle is the bisector of the angle. (See **Figure 3**.)

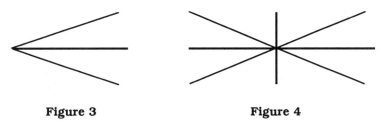

Figure 3   Figure 4

Theorem L.4: The locus of a point equidistant from two given intersecting lines is the bisectors of the angles formed by the lines. (See **Figure 4**.)

Theorem L.5: The locus of a point equidistant from two concentric circles is the circle concentric with the given circles and midway between them. (See **Figure 5**.)

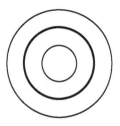
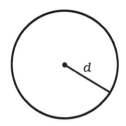

Figure 5   Figure 6

Theorem L.6: The locus of a point at a given distance from a given point is a circle whose center is the given point and whose radius is the given distance. (See **Figure 6**.)

Theorem L.7: The locus of a point at a given distance from a given line is a pair of lines, parallel to the given line and at the given distance from the given line. (See **Figure 7**.)

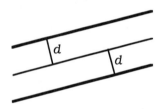
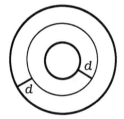

Figure 7   Figure 8

Theorem L.8: The locus of a point at a given distance from a given circle whose radius is greater than that distance is a pair of concentric circles, one on either side of the given circle and at the given distance from it. (See **Figure 8**.)

Theorem L.9: The locus of a point at a given distance from a given circle whose radius is less than that distance is a circle outside the given circle, at the given distance from it, and concentric to it. (See **Figure 9**.)

Figure 9   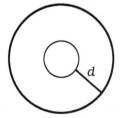

344   APPENDIX 4

Appendix 5

Some Area and Perimeter Formulas

Definition: A **square unit** is the surface enclosed by a square each of whose sides is 1 unit.

Definition: The **area of the interior of a closed plane figure** is the number of square units contained in its surface.

Area Formula 1: For a rectangle, the area of its interior is given by $A = LW$, where L is the length and W is the width.

Area Formula 2: For a triangle, the area of its interior is given by $A = \frac{1}{2}BH$, where B is the length of one side and H is the length of the altitude to that side from the vertex opposite that side.

Area Formula 3: (**Heron's Formula**) For a triangle, the area of its interior is given by $A = \sqrt{S(S-a)(S-b)(S-c)}$, where a, b, c are the lengths of the sides of the triangle and $S = \frac{a+b+c}{2}$ is the semi-perimeter of the triangle.

Area Formula 4: For a parallelogram, the area of its interior is given by $A = BH$, where B is the length of one side and H is the length of an altitude to that side from a vertex opposite that side.

Area Formula 5: For a trapezoid, the area of its interior is given by $A = \frac{1}{2}H(B+B')$, where B and B' are the lengths of the parallel sides and H is the perpendicular distance between these sides.

Area Formula 6: For a circle, the area of its interior is given by $A = \pi R^2$, where R is the radius of the circle.

Definition: A **sector of a circle** is the part of the interior of the circle which is bound by two radii and their intercepted arc.

Definition: A **segment of a circle** is the part of the interior of the circle which is bound by a chord and its arc.

Area Formula 7: For a circle of radius R, the area of a sector whose central angle measures n degrees is given by $A = \frac{n}{360}\pi R^2$.

Area Formula 8: For a circle of radius R, the area of a sector whose central angle measures t radians is given by $A = \frac{1}{2\pi}\pi R^2 = \frac{1}{2}tR^2$.

Definition: The **perimeter** of a simple closed plane figure is the length of the segment which would result if the figure were cut at one place and laid out along a line.

Perimeter Formula 1: For a polygon, its perimeter is the sum of the lengths of its different sides.

Perimeter Formula 2: For a circle, its perimeter, or **circumference**, is given by $C = 2\pi R = \pi D$, where R is the radius of the circle and D is its diameter.

Perimeter Formula 3: For a circle of radius R, the length of an arc whose central angle measures n degrees is given by $S = \frac{n}{360}2\pi R = \frac{n}{180}\pi R$.

Perimeter Formula 4: For a circle of radius R, the length of an arc whose central angle measures t radians is given by $S = \frac{1}{2\pi}2\pi R = tR$.

BIBLIOGRAPHY

ADLER, CLAIRE FISHER. *Modern Geometry*. New York: McGraw-Hill, 1958.
AVERY, ROYAL A. and WILLIAM C. STONE. *Plane Geometry*. Boston: Allyn and Bacon, 1961.
BIRKHOFF, GEORGE DAVID and RALPH BEATLEY. *Basic Geometry*. 3rd ed. New York: Chelsea, 1959.
BORSUK, KAROL and WANDA SZMIELEW. *Foundations of Geometry*. Amsterdam: North-Holland, 1960.
BROWN, RICHARD G. *Transformational Geometry*. Lexington, MA: Ginn, 1973.
COPI, IRVING M. *Symbolic Logic*. New York: MacMillan, 1954.
GANS, DAVID. *An Introduction to Non-Euclidean Geometry*. New York: Academic Press, 1973.
GOLOS, ELLERY B. *Foundations of Euclidean and Non-Euclidean Geometry*. New York: Holt, Rinehart and Winston, 1968.
GREENBERG, MARVIN JAY. *Euclidean and Non-Euclidean Geometries*. San Francisco: Freeman, 1974.
HILBERT, DAVID. *The Foundations of Geometry*. 1st English ed. La Salle, IL: Open Court, 1902.
HILBERT, DAVID. *The Foundations of Geometry*. 2nd English ed. La Salle, IL: Open Court, 1971.
HIRSCH, CHRISTIAN R., MARY ANN ROBERTS, DWIGHT O. COBLENTZ, ANDREW J. SAMIDE, and HAROLD L. SHOEN. *Geometry*. Glenview, IL: Scott, Foresman, 1979.
MARTIN, GEORGE E. *The Foundations of Geometry and the Non-Euclidean Plane*. New York: Springer-Verlag, 1975.
MARTIN, GEORGE E. *Transformation Geometry*. New York: Springer-Verlag, 1982.
MESERVE, BRUCE E. and JOSEPH A. IZZO. *Fundamentals of Geometry*. Reading, MA: Addison-Wesley, 1969.
MILLER, LESLIE H. *College Geometry*. New York: Appleton-Century-Crofts, 1957.
MOISE, EDWIN E. *Elementary Geometry from an Advanced Standpoint*. Reading, MA: Addison-Wesley, 1963.
PRENOWITZ, WALTER, and MEYER JORDAN. *Basic Concepts of Geometry*. New York: Blaisdell, 1965.
RINGENBERG, LAWRENCE A. *College Geometry*. New York: John Wiley & Sons, 1968.
SHIVELY, LEVI S. *An Introduction to Modern Geometry*. New York: John Wiley & Sons, 1939.
SMART, JAMES R. *Modern Geometries*. 3rd ed. Pacific Grove, CA: Brooks/Cole, 1988.
WILDER, RAYMOND L. *Introduction to the Foundations of Mathematics*. New York: John Wiley & Sons, 1952.
WOLFE, HAROLD E. *Introduction to Non-Euclidean Geometry*. New York: Holt, Rinehart and Winston, 1945.
WYLIE, C. R., Jr. *Foundations of Geometry*. New York: McGraw-Hill, 1964.

INDEX

altitude 198
altitude line 198
angles 107, 278
    acute 144, 173, 253, 296, 302
    adjacent 139
    alternate interior 159, 260
    bisector 169
    comparisons 138, 163
    congruent 126, 295
    consecutive 293
    corresponding 160, 265
    exterior 159, 278
    exterior of 108
    external bisector 237
    included between 108
    interior 159, 283
    interior of 108
    internal bisector 237
    less than 138
    linear pair 139
    obtuse 144
    opposite 107, 293
    quadrilateral 293
    remote exterior 163, 283
    remote interior 163, 283
    right 142
    side 107
    sideline 107
    summit 293
    supplementary 141
    trilateral 278
    vertex 107
    vertical 142
angle of incidence 67
angle of reflection 67
argument 6
    direct 11, 12
    indirect 11, 12
    invalid 11, 12
    patterns 10, 11, 12
    valid 6, 11, 12
Aristotle 5
axiomatic system 3, 50
    complete 59
    satisfiable 51
axioms 3
    independence 50
    independent 57

between 78
    halflines 114, 248
    points 78, 255
biconditional statement 7
bisector
    angle 169
    external 237
    internal 237
    perpendicular 174
    segment 167
Bolyai 74, 247

categorical 60
centroid 202
Ceva 203, 204
circle 197
    center 197
    circumcircle 198
    diameter 197
    incircle 202
    nine-point 217, 219
    orthogonal 241
    radius 197
    secant line 197
    tangent line 197
circumcenter 198
circumcircle 198
claim 26
collinear points 77

comparison
    angles 138, 163
    segments 135, 165, 173, 297
complete 59
completeness 50, 59
    test for 60
compound statement 7
conclusion 4, 6
concurrent lines 196
conditional statement 7
congruence 124
    angles 126
    angle side angle 132
    quadrilaterals 305
    segments 124
    side angle angle 166
    side angle side 129
    side side side 134
    triangles 127
    trilaterals 286
conjecture 3
conjunction 6, 7
consistency 50
    relative 50, 51
    test for 51
contradiction 9
contrapositive 7, 11
contrary 8, 9
converse 7
convex 93, 185, 293
cross ratio 226
    pencil 232
    points 226

deductive reasoning 2, 3, 4
definition 2
denial 7
dilation 191
direct argument 11, 12
disjunction 6, 7
dividing point 204
domain 18

endpoint
    halfline 82
    segment 86
equidistant 174
Euclid 2, 74, 195, 247
Euler line 222
existence theorem 14
exterior
    of angle 108
    of triangle 103
    of trilateral 278
external bisector 237

Fano 43
finite geometry 23
fixed point 150
function composition 150

Gauss 74, 247
geometry
    finite 1, 23
    hyperbolic 195, 247
    linear 76, 181
    plane 76, 181
    space 181
Gergonne 208
glide 117
glide reflection 119, 150

halflines 82
    between 114, 248
    endpoint 82
    opposite 82
    opposite sides of 82
    parallel 159, 247
    perpendicular 169
    same side of 82
    side of 90
halfplanes 90
    opposite 90
    opposite sides of 90
    same side of 90

halfspaces 184
    opposite 185
halfturn 97
harmonic
    conjugates 235
    mean 238
    pencil 235
    points 234
    range 235
Hilbert 1, 74, 195, 247
hyperbolic
    geometry 195, 247
    parallel axiom 247

image 18, 230
implication 7
incenter 201
incircle 202
independence 50
    axioms 57
    test for 57
indirect argument 11, 12
inductive reasoning 2, 3, 4
interior
    of angle 108
    of triangle 102
    of trilateral 278
internal bisector 237
interpretation 51
invalid argument 11, 12
inverse 7
isometry 46, 149
isomorphic 60

Klein 74, 247

Lambert 74, 302
language rule 23
Law of Contradiction 6
Law of Excluded Middle 6
less than
    angles 138

    segments 135
line reflection 64, 150
lines
    concurrent 196
    coplanar 182
    Euler 222
    opposite sides of 90
    parallel 29, 159, 247
    pencil 231
    perpendicular 170
    same side of 90
    secant 197
    side of 90
    Simson 217
    space 183
    symmetry 176
    tangent 197
Lobachevsky 74, 247
logic 5
logically equivalent 7
logical reasoning 4
logical subsystem 52
logical system 3
logic forms 6

mapping 17
    one-to-one 18
    onto 18
mean
    harmonic 238
    proportional 240
median 202
median line 202
Menelaus 203, 209
midpoint 167
model 51

negation 6, 7
nine-point circle 217, 219
notation rule 23

one-to-one correspondence 59

orientation 70, 100
orthocenter 200
orthogonal circles 241

parallel halflines 247, 265
parallel lines 29, 159
    alternate interior angles 159
    corresponding angles 160
    exterior angles 159
    hyperbolic 247
    interior angles 159
    transversal 159
partition 83
    line 83
    plane 90
    segment 87
    space 185
Pasch 74, 87
pencil
    cross ratio 232
    harmonic 235
    lines 231
    vertex 232
perpendicular
    bisector 174
    halflines 169, 250
    lines 170
    segments 170
planes
    opposite side of 185
    partition 90
    same side of 185
    side of 185
points
    between 78, 255
    collinear 77
    concurrency 196
    coplanar 182
    cross ratio 226
    dividing 204
    fixed 150
    harmonic 234

perpendicularity 169
range 231
space 183
symmetry 178
postulate 3
preimage 18
premise 4, 6
preserve relations 60
primitive term 2, 56
principle of mathematical
    induction 93
projection 20, 230
    center of 230
    central 230
    image 230
proof strategy 23
proportional mean 240
proposition 6

quadrilateral 292
    adjacent sides 293
    angle 293
    base 293
    congruent 305
    consecutive vertices 293
    convex 293
    diagonal 293, 299
    Lambert 302
    opposite sides 293
    Saccheri 293
    side 292
    sideline 292
    summit 293
    vertex 292
quantifier 8
    existential 8
    universal 8, 9

range 18
range of points 231
    base 231
    harmonic 235

ratio
    division 204
    similarity 188
    stretch 188
reflection 64
    angle of 67
    line 64, 150
rotation 96, 150
    center of 96
    identity 97

Saccheri 74, 293
satisfiable system 51
secant line 197
segment 86
    comparison 135, 165, 297
    congruence 124
    directed 203
    dividing point 204
    endpoints 86
    length 203
    less than 135, 297
    measure 203
    midpoint 167
    parallel 159
    side of 90
separate 84
    line 85
    plane 90
    segment 87
    space 185
side
    adjacent 293
    included between 108
    of a quadrilateral 292
    of a triangle 102
    of a trilateral 278
    of an angle 107
    opposite 102, 107, 279, 293
similarity 188
    ratio of 188
simple statement 7

Simson 215
Simson line 217
space 181
    line 183
    point 183
statement 6
    biconditional 7
    compound 7
    conditional 7
    contradictory 7
    contrary 8, 9
    negation 6, 7
    simple 7
strategy 23
stretch 188
    center of 189
    identity 189
    ratio 188
    reflection 189
    rotation 190
subsystem 52
symmetry 178
    identity 177
    line of 176
    point of 178
    rotational 177

tangent line 197
test for completeness 60
test for consistency 51
test for independence 57
theorem 3
    existence 14
transformation 19
    identity 22, 189
    translation 117, 150
        identity 117
    transversal 159
    triangle 102
        altitude 198
        altitude line 198
        angle 107

INDEX 353

centroid 202
circumcenter 198
circumcircle 198
congruent 127
exterior 103
exterior angle 163
incenter 201
incircle 202
interior 102
interior angle 163
isosceles 129
median 202
median line 202
nine-point circle 217, 219
opposite angle 107
opposite side 102
opposite vertex 102
orthocenter 200
remote exterior angle 163
remote interior angle 163
side 102
sideline 102
vertex 102
trilaterals 277
    angle 278
    congruence 286
    exterior 278
    exterior angle 278
    inner side 278
    interior 278
    interior angle 283
    isosceles 288
    opposite outer side 279
    opposite vertex 279
    outer side 278
    remote exterior angle 283
    remote interior angle 283
    side 278
    vertex 278
truth table 6
truth value 6

undefined term 2

valid argument 6, 11, 12
vertex
    consecutive 293
    of angle 107
    of pencil 232
    of quadrilateral 292
    of triangle 102
    of trilateral 278
    opposite 102, 279, 293

Young 1